Out of Time

Out of Time

A Philosophical Study of Timelessness

SAM BARON, KRISTIE MILLER, AND
JONATHAN TALLANT

OXFORD
UNIVERSITY PRESS

OXFORD
UNIVERSITY PRESS

Great Clarendon Street, Oxford, OX2 6DP,
United Kingdom

Oxford University Press is a department of the University of Oxford.
It furthers the University's objective of excellence in research, scholarship,
and education by publishing worldwide. Oxford is a registered trade mark of
Oxford University Press in the UK and in certain other countries

© Sam Baron, Kristie Miller, and Jonathan Tallant 2022

The moral rights of the authors have been asserted

First Edition published in 2022

Impression: 3

Published in the United States of America by Oxford University Press
198 Madison Avenue, New York, NY 10016, United States of America

British Library Cataloguing in Publication Data
Data available

Library of Congress Control Number: 2021952608

ISBN 978–0–19–286488–8

DOI: 10.1093/oso/9780192864888.001.0001

Printed and bound in Great Britain by
Clays Ltd, Elcograf S.p.A.

Table of Contents

List of Figures and Tables

Figures

Tables

Introduction

Why write a book about timelessness? After all, it appears *obvious* that there *is* time. The contention that time does not exist appears transparently and straightforwardly undermined by just about every experience that we have. To be sure, there is scope to debate exactly what time is like, but not whether it exists. On the contrary, we think that time might not exist. Our goal in this book is to convince you of the same.

We focus on a very specific conception of time, namely: the everyday or 'folk' concept of time. Time, in this sense, might not exist. The 'might' here is fairly weak: it is epistemically possible that nothing satisfies the folk notion of time. In this way our project is similar to an influential view about colour. According to colour anti-realism, colours in the folk sense—i.e., intrinsic properties that inhere in objects—do not exist. We won't defend an analogous temporal anti-realism. We only hope to show that it could be true.

That's hard enough. The loss of time in the folk sense would seem to have widespread implications for our everyday lives. The most obvious implication concerns agency. Agency appears deeply connected to time. Moral and prudential planning both seem to make sense only in the presence of a future to plan for, and a past on the basis of which to make such plans. And so, if time does not exist, then a large-scale reconsideration of these notions would seem in order—if, indeed, anything resembling what we consider to be agency or planning could survive within a timeless reality.

Given that we all know we are agents, this forms the basis of a powerful argument against the possibility that time does not exist. The argument has a Cartesian flavour: we know for certain that we are agents, time is needed for agency, so we know that time exists. Ultimately, we aim to show that agency can exist even though time, in the folk sense, does not. Thus, the considerations surrounding agency that loom so large in our lives do not provide a decisive case in favour of time's existence.

We get to this conclusion by defeating two further reasons to suppose that time, in the folk sense, must exist. The first of these relates to the concept of time. The thought is that the everyday notion of time is immune to error, in this sense: no matter what we discover about the world people will continue to

Out of Time: A Philosophical Study of Timelessness. Sam Baron, Kristie Miller, and Jonathan Tallant, Oxford University Press.
© Sam Baron, Kristie Miller, and Jonathan Tallant 2022. DOI: 10.1093/oso/9780192864888.001.0001

judge that time exists. Here we analyse the results of some of the first empirical work on folk notions of time ever conducted. On the basis of this work, we argue that there is evidence that certain discoveries about the world would lead people to conclude that time does not exist.

Of course, even though there are some hypothetical discoveries that would lead people to deny that time exists, it doesn't follow that such discoveries could really be made. Thus, one might continue to believe that folk time must exist on the grounds that the discoveries that would lead people to reject the existence of time are incompatible with what we know about the world, based on our best science. We show that this is not so, by taking a look at some recent developments in quantum gravity.

This allows us to finally tackle agency. Using a notion of causation that has been cleaved from the folk concept of time, we show how to rebuild agency. In this way, we show that agency might exist despite the absence of time in the folk sense. As a corollary, we show that the loss of time is no big deal. We can continue to live our lives much as we ever did even if we discover that time does not exist. This sets the scene for a philosophical case against the existence of time. For if we don't need time to live our lives, then what, in the end, is it good for? Perhaps time, like colour before it, is a metaphysical dangler; a notion that we use to make sense of our lives, but not one that deserves a place in a considered ontology of the world.

PART I

1

Folk Concepts of Time

1.1. Introduction

The idea that time does not exist is, for many, unthinkable: time *must* exist. Our goal is to make the absence of time thinkable. Time might not exist. This chapter lays the groundwork for our investigation. We begin by clarifying the central target of our investigation, the folk concept of time and then motivating the idea that it is this folk concept whose investigation matters (§1.2). We then use this to identify three respects in which the loss of time might be deemed unthinkable, which we aim to address in the book (§1.3). After that, we head off some potential misconceptions about the project (§1.4), before saying a bit about the methodology that we employ (§1.5). Finally, we provide an overview of the three main parts of the book (§1.6). This will be deliberately programmatic, to be filled out in later chapters.

1.2. Motivation

Our primary focus in this book is on the folk concept,[1] or concepts, of time. We will have much more to say about how we think about folk concepts shortly. But for now, we can think of the folk concept of time as *something like* the naïve view of time of the sort introduced by Callender (2017): the unreflective notion of time that individuals use in their everyday lives. To be clear, that's not *quite* right. When we talk of the folk concept of time we don't simply mean the way the folk think about, or conceive of time. We don't simply mean what the folk think time is like. We mean something like *what the folk think (almost certainly implicitly) it would take for there to be time in a world.*

For the time being we will suppose that there is a single such folk concept, though as we shall see later on, matters are much more complex. When we make judgements about there being certain temporal relations, such as judging

---

[1] A concept employed by non-philosophers.

Out of Time: A Philosophical Study of Timelessness. Sam Baron, Kristie Miller, and Jonathan Tallant, Oxford University Press.
© Sam Baron, Kristie Miller, and Jonathan Tallant 2022. DOI: 10.1093/oso/9780192864888.003.0001

that E is earlier than E*, or that E* is tomorrow and not yesterday, or that X and Y occur at the same time, these judgments are all intimately connected with our folk concept of time. When we judge that time has a direction, or that time flows, (if we do) these are judgements that are intimately connected with our ordinary concept of time.

Why should we care about the folk concept of time? The short answer is that we should care about time in the folk sense, because it appears to be implicated in normative concepts and practices in which we are deeply invested; concepts and practices like moral and practical responsibility.[2] We take there to be a tight connection between (certain) concepts and practices. For instance, we take there to be a tight connection between various normative concepts, such as the concept of moral responsibility or the concept of prudential rationality, and certain practices, such as the practices of holding people morally or practically responsible. We will suppose that these practices are ones that rely on deploying the concepts in question. We can engage in the practice of holding someone morally responsible only because we have and deploy the concept of moral responsibility.

Notably, then, some of our normative practices—both moral and prudential—appear to involve deploying not only normative concepts, but also temporal concepts. Moral and practical reasoning apparently involves our taking the ways things are at some times, to be moral or practical reasons for action at some other time.

The loss of time in this sense thus seems to threaten our capacity to reason coherently as moral and prudential agents and, in this way, promises to rob of us of agency. This last point may not seem obvious, and so it is worth saying a bit more about the threat posed to agency by the loss of time in the folk sense. After all, science often shows our folk concepts to be in error and this doesn't usually make much of a difference to the life of the average person. Consider, for instance, the case of colour. Colour anti-realism of some kind appears to be vindicated by science. Colour science seems to reveal that there is nothing in the world that answers to the folk concept of colour. The folk concept of colour (let us suppose) treats colours as monadic, intrinsic properties of objects and,

---

[2] Of course, we recognize that the distinction between scientific and folk concepts of time is somewhat artificial. The scientific concept is surely at least in part influenced by the folk concept, in that the folk concept guides theorising about what sort of thing in the world is even a candidate to answer to the scientific concept. Scientists look for time in certain places, and not others, because, at least at the beginning of investigation, they use the folk concept of time to determine what sort of thing they are attempting to locate. The converse is also the case. The scientific concept is likely to influence the folk concept (and, we suspect, already has, since the rise of relativity). One of the tasks we set ourselves in this book is to say something about the content of the folk concept.

supposedly, there are no such properties (Palmer (1999, 95), Zeki (1983), Land (1983), and Kuehni, Rolf and Hardin (2010)). Nonetheless, this doesn't seem to impact on the folk in any significant fashion.

There is, however, a key difference between the colour and temporal cases. Colour *experience* is left intact by science. Science does not tell us that our experiences of colours are not real, it just tells us that our experiences of colours are not experiences of intrinsic colour properties. What we are experiencing is rather unlike the way we conceptualise what we are experiencing.[3] And, because our colour-based behaviours and practices are all based on colour experiences, there is no problem with continuing to engage in these practices despite the discovery that our folk concept of colour is in error. No colour, no problem.

In contrast, the practices that are built up around the concept of time are *not* based entirely on our temporal experiences. Rather, they are based on *beliefs* that time, under a certain conception, exists. Such beliefs include the belief that there is a past and future in some sense. Such beliefs appear to be tightly related to the folk concept of time and, in turn, to practices surrounding decision-making, which are integral to our self-conceptions as agents.[4]

Unlike the practices surrounding colour, these practices seem to demand that the world be a certain way. In particular, it cannot merely seem to us as though we have agency (just as it may seem to us that there are colours). There must really be agency in order for the practices surrounding decision-making to be in good standing. While we cannot provide a full-blown analysis of the folk concept of agency, we can take the sorts of activities that are associated with agency as a guide in helping us to explore this point. Agency, we will assume, is associated with decision-making practices of at least two kinds: prudential and moral. First: prudential practices.

---

[3] To be clear, we are not saying that colour plays no important role in the lives of the folk. Clearly, there are a great many colour-based practices about which we care deeply. Practices like art, design, symbolism, advertising, and the like. The point, rather, is that all of these practices can be scaffolded by the experiences alone. The reality of colour, as we conceptualise it at an everyday level, is not necessary for those practices to be in good standing.

[4] Suppose we manage to recover our experiences intact, despite recognising that time in the folk sense doesn't exist. We might say that all is well, because we have the appearance of agency even though no one really has it. But what is the appearance of agency? Presumably, this would be something like the following: it seems as though one can exercise one's agency when, in fact, one cannot. So it seems as though one can act now to bring about an outcome in the future when, in fact, this is false, because the world does not answer to the sense of time that underwrites the related notion of 'future'. The mere appearance of agency is all but useless, we contend. Insofar as one desires agency, one does not desire for it to appear as if one can make choices, knowing full well that one cannot. One desires to *really be able* to make choices. So while colour experience in the absence of real colours can underpin practices of colour discrimination, design, preference, and so on, the experience of agency cannot underpin agency in the absence of real agentive outcomes.

Prudential planning is the kind of planning one undertakes to bring about one's desires given one's beliefs. Agents in this mode consider their options, and make a choice based on the information that they have available to them. Prudential planning is always planning for what we believe to be in the future, and never for the past.[5] We take certain events to be fixed in time, then, based on those fixtures, consider our options, subsequently (all going well) selecting the best—i.e., the one that is most likely to bring about a desired outcome. This practice of prudential planning makes sense only if we believe there is a future to plan for, and if there is past information that is relevant to constraining one's future plans. Without any past information, any decision is as good as any other, since as far as one can tell all decisions are equally likely to bring about one's desired outcomes.[6]

Time is also implicated in moral practice—in two ways. First, consider moral responsibility. We commonly hold people to be morally responsible for acts performed in the past. Suppose we then discover that there is no such thing as the past and so come to form just such a belief. Then, in a certain sense, it is just *not true* that anyone did anything wrong, because it is not true that anyone *did* anything. Accordingly, without a past to believe in, we will not believe that anyone *did* anything wrong. Second, time also seems to be implicated in the practice of moral reasoning. Moral reasoning involves working out what one ought to do, morally speaking. Making a decision about what one ought to do is making a decision about future outcomes; a decision based on information about the past. If there is no past and no future to believe in, then deciding what one ought to do in the future in a moral sense is simply not a meaningful notion.

Perhaps without time in the folk sense we would retain the ability to act as though we are agents. For instance, if we have false beliefs about the past, then maybe we could in some sense exercise our capacities for both moral and prudential reasoning. Arguably, however, this would not be to engage in genuine moral or prudential reasoning; it would certainly not involve being a *genuine* agent. Moreover, insofar as we come to know that there is no time,

---

[5] Setting aside recherché cases of backwards time travel. Since there is, as a matter of fact, no such time travel (granting that it is physically and logically possible), planning is, in fact, always for the future.

[6] Perhaps there can be non-temporal notions of agency. That is something we need to think about very carefully. The point, however, is that the everyday notion of agency, the one that is associated with planning behaviour, appears to be tightly connected to our everyday thinking about time. We can frame the matter in terms of a conceptual web. At the centre of the web, or close to the centre, is time. Agency and morality bear substantial connections to time, but are less central to the web. When we lose time, we must refigure those concepts in the web that are connected to the time concept but are more peripheral, or else give up on those concepts as well.

it's hard to see why anyone would *bother* exercising their apparent agency. If there is no future in the sense needed for everyday moral and prudential reasoning, what would be the point? There is nothing to be planned for and no basis for any decision.

In the absence of moral and prudential decision-making practices, it would seem to follow, in quick order, that there is no genuine way to be an agent. It is tempting to think that, in such a situation, there simply are no agents in the folk sense.[7] This loss of time in the folk sense thus represents a serious threat of agentive paralysis. If nothing in the world answers to the notion of time that underwrites agency, then it is unclear that we could, would, or should engage in agentive behaviour.

In sum, then, the loss of time in the folk sense threatens our capacity to reason coherently as moral and prudential agents and, in this way, promises to rob of us of agency. We should thus care about the folk concept of time because it is connected to agentive notions that we, the folk, care about in a manner that is not clearly true for some scientific or philosophical conception of time.

1.3. The Unthinkability of Timelessness

Agency not only gives us a reason to care about the folk concept of time, it also provides one of the core reasons to suppose that time in this sense must exist, and thus that the loss of time is unthinkable. After all, if time in the folk sense is required for agency, and we know that we are agents, then it appears we are forced to conclude that time in the relevant sense exists.

Ultimately, we aim to argue against this claim: time is not needed for agency. In order to get there, however, we need to consider two further reasons why one might take it to be unthinkable that time, in the folk sense, does not exist. First, one might worry that time in the folk sense is simply immune to error. Basically, there are no discoveries that one could make about the world that would lead people to conclude that time does not exist. The hypothesis of immunity seems plausible in light of the deep connection between time and agency. If agency requires time, and people self-conceive as agents, then we might well expect that people simply won't judge that time doesn't exist no matter what you tell them about the world. We argue against this claim too.

---

[7] But such a strong conclusion is not strictly needed. Even if the right thing to say is that we remain agents but can't exercise our agency in any meaningful way, that would be a substantial blow.

We show that there are, in fact, situations in which people will judge that time does not exist when presented with certain discoveries about the world. This begins to drive a wedge between time and agency.[8]

Next, we turn to science. One might concede that there are some hypothetical discoveries that would lead people to conclude that time does not exist, but nonetheless maintain that such discoveries are incompatible with what we know about the world. The loss of time in the folk sense is thus unthinkable because it is epistemically impossible given the implications of science. Again, this might seem quite plausible, especially in light of physics. According to the general theory of relativity, spacetime is a basic constituent of reality.

However, we argue that recent developments in physics present a serious challenge to the existence of spacetime in at least *some* sense. It thus seems hard to maintain that there are no hypothetical discoveries that we could make that would lead people to conclude that time does not exist.

Next we argue that causation and the folk notion of time come apart. This sets the scene for our return to agency. Because the folk notion of time and causation come apart, it is possible to have agency in the absence of time in the folk sense. We can use causation in the absence of time as a new foundation for agency. In this way, we show that agency provides no reason to suppose that time, in the folk sense, must exist.

All in all, then, we make a case against three 'unthinkabilities' of timelessness by showing (i) that the folk concept of time is not immune to error; (ii) the discoveries that would undermine the folk concept of time are not incompatible with what we know about the world; and (iii) time is not needed for agency. We stop short of the conclusion that time, in the folk sense, does not exist. But what we say does set up the conditions for an argument against time in the relevant sense. If we don't need time for agency, then it is entirely possible for developments in physics to eliminate time completely. Time, in the everyday sense, is a negotiable feature of reality.

1.4. Misconceptions Addressed

Before we go any further, it is important to address five potential misconceptions about our project. First and foremost, we do not, in any way, take the folk

---

[8] We recognise the possibility that people won't in fact respond in the ways they say they will, and thus that certain discoveries won't lead to widespread abandonment of the folk concept of time. We flag this as an area for further empirical work, and proceed under the assumption that things won't work out this way.

understanding of time to be evidence for a particular picture of reality. Nor are we taking intuitions about time as evidence that time is thus and so.[9] At most, we take folk intuitions to provide partial evidence about the content of the folk concept, or concepts, of time.

Second, we are not interested in criticising science using a folk concept. Thus, we are not offering anything like the following argument: our best physical theories imply that there is nothing that satisfies the folk concept of time; if the folk concept of time is unsatisfied, then there is no such thing as agency; there clearly is agency, and so our best physical theories are wrong. We don't endorse this argument and we don't endorse the idea that folk notions of time might provide a basis for rejecting a physical theory. To put the point in more general terms, we are not engaged in *first philosophy* (the idea, roughly, that philosophy comes first, and science after, thereby enabling criticisms of scientific theory on purely philosophical grounds).

Third, while we are not interested in constraining science with the folk concept, we are not willing to give physics a free ride either. We don't believe that one should uncritically accept any interpretation of a physical theory, even one provided by a physicist who is engaged in interpreting their own theory. We maintain that the project of interpreting a scientific theory is a substantive and often philosophical project, and in the process of working out what a scientific theory tells us about reality, philosophy can often play a useful role. Insofar as we appear to be criticising a scientific theory, we take ourselves to be criticising a particular, philosophically laden interpretation of that theory.

Fourth, we are not presuming that there is a single folk concept of time. Nor are we assuming that the folk concept of time is some specific way. Nor are we willing to simply introspect on our concepts and just go with whatever seems right to us, using it as a proxy for what the folk (tacitly) believe about time. The content of the folk concept is something we have to figure out, and it is not something that can be determined by three philosophers introspecting on their own concepts. We are very much open to the idea that there are many folk concepts of time that vary from person to person, and that may be messy, indistinct and incomplete.

Finally, we are not presuming that there is a sharp distinction between folk and scientific concepts of time. It may be that these concepts overlap in various

---

[9] Perhaps there could be some role for folk intuitions about time in choosing between empirically consistent theories: namely we should choose the theory that best accords with the folk concept of time. We (collectively) are not even sure this is an appropriate use of such intuitions and we won't appeal to such considerations in this book.

ways. We are also not assuming that these are the only concepts of time. Metaphysics, history, economics, literature, and more may all work with different notions of time. It may even be that, within science itself, there are multiple notions of time at work.

1.5. Methodology

In line with our rejection of first philosophy, our methodology in this book is broadly naturalistic in flavour. Obviously, this means that we take science seriously and as a touchstone for any metaphysical investigation. We also carry this naturalistic approach through to the conceptual analysis needed to get a grip on the folk concept of time.

The standard methodology of conceptual analysis is roughly conceptual analysis on the Canberra plan.[10] It involves identifying a range of platitudes about the concept of interest—in this case, time—and then using those platitudes as a basis for analysing that concept. So, for instance, in order to provide a conceptual analysis of *pain* we might identify a range of folk platitudes about pain. We then systematise these and, at the end, come up with a list of necessary and sufficient conditions for something to be pain. That is our candidate conceptual analysis. Then we look to the world to see what, if anything, satisfies that analysis; if something does, then that thing is pain.

In the last twenty years or so, we have seen the rise of experimental philosophy as an alternative to conceptual analysis on the Canberra plan. Rather than philosophers sitting in their offices, gathering folk platitudes via introspection (or maybe asking their friends) and then regimenting them with a theory, the suggestion is that we should find ways to experimentally probe the folk in order to reveal the concepts that they use. The early success of experimental philosophy was simultaneously the introduction of a new method of conceptual analysis and a criticism of the old. We now have good evidence coming from experimental philosophy that there is substantial cross-cultural and gendered variation in the folk concepts that have been traditionally subject to philosophical analysis.[11] Given this variation, we must be very careful

---

[10] See Braddon-Mitchell and Nola (2008) for discussion of this approach.

[11] For instance see Weinberg, Nichols and Stich (2001), Machery, Mallon, Nichols, and Stich (2004), Buckwalter and Stich (2014), and Friesdorf, Conway and Gawronski (2015). The extent to which these differences exist is controversial, as a number of such studies have failed to replicate. See for instance Nagel, San Juan and Mar (2013), Kim and Yuan (2015), and Adleberg, Thompson and Nahmias (2015).

about generalising any Canberra-plan style analysis of a concept beyond the cultural milieu in which the philosophers putting together the analysis are situated.[12]

Our own exploration of the folk concept of time is empirical in nature. We do not analyse the folk concept of time by starting with claims that are (perceived to be) platitudes about time and then regimenting them into an analysis. Rather, we draw on a series of studies performed by one of the authors (Kristie Miller, and her team at Sydney) that seek to examine the structure of folk temporal concepts. These studies are informed by the platitudes that are the bread and butter of traditional conceptual analyses, and by the metaphysical theories of time that philosophers have developed. After all, empirical work requires hypotheses, and we use what philosophers have said about time to generate some. The studies are survey-based, and involve presenting participants with vignettes about various ways the world could be. Agreement is then sought about whether time exists in the relevant situations. The idea is to try and find cases in which participants will generally agree that time exists, and cases in which they will generally disagree that this is so. We can then use the pattern of agreement and disagreement to test various hypotheses about the content of any folk temporal concepts.

Over the course of the next three chapters, we will describe the outcome of this empirical work. However, it is worth emphasising that we don't attempt to present a conceptual analysis of the folk concept (or concepts) of time on the basis of these results. That is, we won't be presenting a set of necessary and sufficient conditions for something to fall under one (or more) folk concepts of time. That is an important, and large, task, but one that outstrips the resources we have. Many more studies would be needed to allow us to generate such an analysis, even assuming such an analysis were possible (something which we consider in Chapter Two).

1.6. Overview

Having said a bit about why we are focusing on the folk concept(s) of time, and a bit about our methodology, we are now in a position to give a broad overview

---

In addition to the failures of replication, there has also be positive evidence for some cross-cultural uniformity at least as regards certain intuitions (see Machery, Stich, Rose, Chatterjee, Karasawa, Struchiner, Sirker, Usui, and Hashimoto 2015).

[12] None of this is to say that we do experimental philosophy in a vacuum. Any hypotheses about the content of a folk concept are surely informed by *something like* the Canberra Plan approach.

of the book. Our central claim is that time, in the folk sense, might not exist. The relevant 'might' is an epistemic might: the non-existence of time is compatible with what we know about the world. But, to be clear, we are not arguing for the claim that time in the folk sense does not exist, nor for the claim that this is a metaphysical possibility. Our claim is more might-y.

The three parts of the book are devoted to repudiating the three reasons why one might think that time must exist. We take these to be that: (i) there are no discoveries we could make about the world that would lead people to judge that time does not exist; (ii) discoveries that would lead people to judge that time does not exist are incompatible with what we know about the world from science; and (iii) time is required for agency.

Part One takes up the challenge of exploring our folk concept (or concepts) of time via empirical investigation. Because we recognise that many readers will be keen to hear the upshot of the empirical work conducted, but may be less interested in the precise details of that work, we present the results of the work straight away, in Chapter Two. There we reflect on what these empirical results tell us about the conditions under which our folk concept of time will be satisfied. We provide evidence against the immunity of folk concepts of time to error. We show that there is a range of discoveries about the world that would lead people to judge that time does not exist. We briefly consider the idea that the folk would continue to say that time exists, despite making the kinds of discoveries that we identify in Part One. We suggest that this is an empirical bet about the world, and one that we don't have evidence for (and some evidence against). Still, we recognise that such bets might well pan out, and thus concede that everything we say is open to refutation via future empirical work. Such is the nature of empirical hypotheses.

The next two chapters are given over to laying out, in detail, the empirical methodology used (Chapter Three), and the nature of the studies conducted (Chapter Four). These chapters are optional chapters for those who want a more detailed picture of the empirical work. In the spirit of a choose-your-own-adventure novel, the reader can skip these empirical chapters, and proceed straight to Part Two of the book, should they wish to do so.

In Part Two, we turn to recent developments in physics. We argue that the kinds of discoveries that would lead folk to deny that time exists may be suggested by our current physics. We base this argument on work in quantum gravity. A number of approaches to quantum gravity suggest that spacetime does not exist, and that the world has the kind of structure that would undermine the folk concept of time. We formulate this as the quick argument for timelessness, which seeks to establish the epistemic possibility of timelessness in the face of scientific knowledge. The quick argument for timelessness faces an immediate 'emergence challenge'. The challenge, in brief, is that spacetime is expected to be an emergent phenomenon. If spacetime is emergent then the necessary conditions needed for time to exist are plausibly in place and so it becomes plausible, once more, that the discoveries that would undermine the folk concept of time are at odds with what we know about the world.[13]

In Chapter Six we provide a response to this 'emergence challenge'. We argue against the idea that spacetime has to be an emergent phenomenon on metaphysical grounds. Roughly put, the problem is that we lack a viable metaphysics of spacetime emergence. We defend this claim by considering a range of different metaphysical conceptions of spacetime emergence, including accounts based on mereological composition and material constitution, grounding, supervenience, and functionalism. In each case we argue that the metaphysical relations used to underwrite the emergence of spacetime either fail to explain how spacetime emerges, or presuppose spatial and temporal notions and so are unfit to connect a non-spatiotemporal reality with a spatiotemporal one.

In Chapter Seven we consider a second tranche of argument that threatens to trump our response to the emergence challenge. The argument is based on a certain problem for theories of quantum gravity: the problem of empirical coherence. The basic thought is that spacetime *must* be emergent for the approaches to quantum gravity that we consider to be open to empirical confirmation and disconfirmation. This is so even if there is no good metaphysics of this style of emergence currently available.

We argue, on the contrary, that the emergence of spacetime is not strictly required to solve the problem of empirical coherence, and sketch a view on which even though spacetime does not exist, the problem of empirical coherence is solved. We note, however, that eliminativists about spacetime nevertheless face a second problem related to observation more generally—called the observation problem—that still needs to be addressed in order to put the emergence challenge to rest.

---

[13] That still leaves open that some of our folk concepts might nonetheless go unsatisfied. If, for instance, our folk concepts were largely A-theoretic in nature, but no actual A-series exists, then this would be the case. This, however, is a threat to the folk concept of time that is not new, and hence is not the threat with which this book concerns itself. It is, however, a threat that our work on the folk concept of time can help to evaluate, since that work will shed light on whether our folk concept of time is indeed A-theoretic.

In the final chapter of Part Two, Chapter Eight, we address the observation problem by outlining an account of causation that is compatible with eliminativism about space-time. We argue, first, that causation can exist even if time does not. Second, we argue that a counterfactual theory of causation can be made compatible with a view on which spacetime fails to emerge. We thus conclude against the emergence challenge and in favour of the quick argument for timelessness. The loss of time in the folk sense remains compatible with what we know from science.

In Part Three of the book, we turn to agency. Our aim is to clarify and defuse the threat to agency that is posed by the loss of time. As we set things up, *temporal error theory* is a view that is analogous to moral error theories. Moral error theories deny the objective reality of moral facts, and, on the basis of this, deny that any moral claims are true. Temporal error theory, then, is the view that temporal thought and discourse is truth apt, and is false (or at least, is not true). In Chapter Nine we argue that temporal error theory is vindicated if our folk concepts of time are not satisfied.

In Chapter Ten, we argue that there is a complex relationship between agentive thought and temporal error theory. We distinguish two versions of temporal error theory: temporal fictionalism and temporal eliminativism. We begin by arguing that temporal eliminativism appears deeply unattractive, and that this appears to militate in favour of temporal fictionalism; the view that we have reason to act as though our temporal thought and talk is true. We then consider whether both temporal eliminativism and temporal fictionalism are in some good sense self-undermining. We suggest that they are if a strong connection between temporal thought and agentive thought is assumed: namely, that if our temporal thoughts are false then so are our agentive thoughts.

In the final chapter, we take up the task of arguing for agentive realism. There, we maintain that the sort of causal structure identified in Chapter Eight can do the job of supporting agency in the absence of time. After explicating this view, we argue that because agentive realism is vindicated, we can make sense of the idea that we can choose to engage in a temporal fiction. Ultimately, we argue that if time does not exist, then we should be temporal fictionalists and agentive realists. This is why the loss of time presents no real threat to us. Our agentive practices remain in good standing since (most of) our agentive thoughts are true (even though our temporal thoughts are false). We have no reason to abandon these practices, or to view them as in some way in error. Granted, our temporal thoughts are not true. Nonetheless, we should continue to engage in our false temporal talk because it is useful in a host of ways. Hence, our temporal practices based around such thought remain as they are.

All of this starts with looking at the folk concept of time. And so, it is to this task that we now turn.

2

Empirical Results

2.1. Introduction

So far as we are aware, there has been no attempt within the philosophical literature to actively study the folk concept of time. To be sure, some philosophers have reflected on their own concepts of time and have encoded these into their metaphysical claims about the nature of time.[1] But, if we wish to know the conditions under which the folk concept of time is satisfied, it's probably better to ask the folk about their concept than to speculate a priori. In fact, it's probably better still to ask them to *use* that concept—for attempting to introspect the application conditions of one's own concepts is a tricky business. Thus, we *won't* be asking the folk about their concept of time (though sometimes psychological research into folk concepts does indeed proceed in this manner); instead, we will be asking the folk to *use* their concept, and we will gather data about how they use the concept across a range of scenarios to inform our understanding of the folk concept(s) of time.[2]

In this chapter, we show that time, in the folk sense, is not immune to error. That is, we show that there are discoveries that one could make about the world that would lead people to deny that time exists. Along the way, we show that there is no single, univocal folk concept of time.

The data discussed here is drawn from a range of empirical studies that we call the Sydney Time Studies.[3] As noted in the introduction, we have structured our discussion of these studies so that much of the technical detail of the

[1] See for instance McTaggart (1908), Zimmerman (2008), Smith (1994), Craig (2000), Schlesinger (1994), Williams (1998; 2003), and Gale (1968). For discussion of the folk concept of time see Baron and Miller (2015a).

[2] By 'concept' we mean a contentful mental state that can be a constituent of a thought. By and large we want to remain neutral with regard to any particular account of concepts. That, of course, can only get us so far. As we will see shortly, we do make some assumptions about what constitutes the content of a concept in order to develop a methodology to probe said concept.

[3] The studies were conducted by Kristie Miller in collaboration with Andrew James Latham and James Norton. We are immensely grateful to both of them for their help with Chapters Three and Four.

Out of Time: A Philosophical Study of Timelessness. Sam Baron, Kristie Miller, and Jonathan Tallant, Oxford University Press.
© Sam Baron, Kristie Miller, and Jonathan Tallant 2022. DOI: 10.1093/oso/9780192864888.003.0002

empirical work can be skipped if necessary (we are conscious that a more philosophically-minded reader may wish to press on). Thus, in this chapter, we report the general findings of the studies and discuss the situations under which 'folk time' goes missing. In Chapter Three we focus on the methodology of the studies, and respond to some worries one might have about how (if at all) it is possible to probe the folk concept of time. In Chapter Four we present the studies in more detail, explaining how they generate the relevant results.

We begin by developing a suite of hypotheses about which conditions are necessary for our folk concept(s) to obtain. To do so, we will look at the sorts of metaphysical models of time that philosophers have proposed, alongside their claims (tacit or explicit) about what is necessary, and what is sufficient, for time to exist. We will present these accounts and then distil from them one or more necessary conditions we can test.

We are principally interested in necessary conditions, since these are what will help us to determine whether there are discoveries that would lead folk to deny that time exists. But because we also want to know whether there is a shared folk concept of time, or multiple distinct concepts, it will be helpful to also look at sufficiency conditions. In order to best speak to this issue, we track the conditions under which people will use their concept of time: the conditions that are sufficient for the concept to be satisfied, as well as the conditions that are necessary for the concept to be satisfied. If we find that there are subgroups in our tested population who disagree about which conditions are sufficient, and which necessary, then this is good evidence that these participants employ different concepts of time. We return to say more about how we can determine whether participants share a concept of time shortly. For now, we just want to note that our aim is to use various philosophical hypotheses, or assumptions, about our concept of time, to develop a series of testable hypotheses.

The chapter is structured as follows. We start, in §2.2, by briefly revisiting some of the standard positions in the philosophy of time. We then go on to use this as a framework for developing a suite of hypotheses about the folk concept of time in §2.3. In §2.4 we summarise the results of the Sydney time studies with respect to the hypotheses identified in §2.3 and discuss how this evidence shows there to be substantial conceptual variation among the folk with respect to their concept of time. In §2.5 we identify a range of discoveries that would lead people to deny that time exists, before briefly considering some responses to the idea that the folk concept of time is immune to error.

2.2. The Metaphysics of Time

We first need some hypotheses about the folk concept of time that we can then test. To locate these, we should consider the traditional debate within philosophy concerning the metaphysics of time. After all, a number of participants to that debate claim to be trying to attack, defend, or otherwise elucidate the 'folk' or 'everyday' concept of time. We can thus use these metaphysical accounts to formulate specific hypotheses about the content of the folk concept(s) of time.

In this section, we will therefore provide a brief reminder of the core positions in the philosophy of time. In the next section, we will use these positions to formulate hypotheses about the folk concept. It is worth emphasising, however, that we view what philosophers have said to date about the folk concept of time to be, at best, a starting point for empirical investigation. As such, some of the hypotheses we ultimately test go beyond the claims that philosophers have made (either directly or indirectly) about what the content of the folk concept of time might be. But we don't see that as a problem. On the contrary, the result is a much richer empirical picture.

Before we get going, some preliminaries. First, in what follows we don't assume that all concepts are explicit. That is, we don't assume that individuals who employ those concepts are always able to *articulate* their content, by providing an analysis of the concept, or providing anything like necessary and sufficient conditions for something to satisfy that concept. Instead, we suppose that much of a person's facility with a concept might be tacit. That is why in our experiments we don't ask people to introspect and then describe the content of their concept. Instead, we just ask them to *use* the concept by judging whether or not certain scenarios are ones in which there is time.

Second, we sometimes talk of 'the' folk concept of time. This, however, is for simplicity: it is not meant to prejudice the issue of whether (a) there is a univocal folk concept *within* the population we experimentally test nor (b) whether there is a univocal folk concept *across* populations. While we cannot speak to (b), we will reflect on (a) in what is to come. For a discussion of what sort of evidence will shed light on the question of whether there is a single shared concept, see in particular Section 4.2.

And so, to the metaphysics. A standard distinction is often drawn between the A-series and the B-series. Both the A-series and the B-series are ways of ordering times or events in such a manner that the ordering has metric structure (i.e., we can talk meaningfully of temporal distances). An A-series ordering orders times in terms of whether they are objectively past, present or future. Importantly, the A-series ordering is a *dynamic* ordering: it is constantly updating with the passage of time. Thus, it may be that a is present, b is past and c is future, but in a moment's time, both a and b may be past and c may be present. The B-series ordering, by contrast, orders times or events in terms of the relations of earlier-than, later-than, and simultaneous-with. The B-series, unlike the A-series, is unchanging. If a is earlier than b, it is always earlier than b. What is earlier than what does not change with the passage of time. Note that when we talk of time passing, or time being dynamical, or temporal passage, we mean temporal passage in the sense defended by A-theorists (and this is what we will always mean by this notion).[4]

In addition to the standard distinction between the A-series and the B-series, we can also differentiate the B-series from the C-series. The C-series, like the A- and B-series, is a way of ordering times and events so as to produce a meaningful notion of temporal distance. Like the B-series, the C-series is static and unchanging. As we will conceive of these series, the central difference between the B-series and the C-series is one of direction.

In the B-series ordering, events are ordered *from* earlier *to* later, and not vice versa. This is to suppose that the B-relations of earlier-than and later-than are directed relations which generate a temporal ordering; on such a picture, there is a fact of the matter regarding which direction is past, and which future. In the C-series ordering, by contrast, events are temporally ordered, but there is no temporal *direction* within the series. One way to think of the C-series is as a betweenness ordering. In a B-series ordering, we might say that a is before b which is before c. In a C-series ordering, by contrast, we say only that b is between a and c. There is no fact of the matter as to whether a is before b and b is before c; or whether c is before b and b is before a.[5] We get metric relations either by counting the number of times between any two times in the ordering (if the ordering is discrete) or via the application of a measure (if the ordering is continuous).[6]

---

[4] This is important, since some B-theorist's use 'temporal passage' to mean something more minimal, which is consistent with a B-theory. See Leininger (2021). On such views one might say that time's having a direction is a product of temporal passage, in this minimal or deflationary sense.

[5] Another useful comparison is that of a hill. Clearly, the top of a hill is higher than the middle and bottom. Nonetheless, the hill is not therefore *height directed*. The hill is not *height-directed* from bottom to top, or top to bottom.

[6] In Chapter Four we also speak of a D-series. While it won't play a direct role in this chapter, it is worth saying a bit here to set up later discussion. In both B- and C-series, there is an ordering. In the B-series that ordering is asymmetric, and in the C-series it is not. In the D-series there is no ordering or metric structure at all. There is just an unordered set of points (i.e. times) in a configuration space that are not metrically related. We set aside the D-series for now and focus on the other three orderings.

Using the A, B and C-series, we can build a range of different theories of time. The three basic views are the A, B and C-theories. According to the basic A-theory of time, events and times are ordered via an A-series. They may or may not also be ordered by a B-series (depending on the particular variety of the A-theory at issue). A-theories of time are sometimes called 'dynamic' theories of time, and we will refer to those who endorse an A-theory as either dynamists or A-theorists (we use this language interchangeably). The point behind the A-series is to capture the idea that the passage of time is a real, objective feature of reality. Temporal passage injects an important dynamic aspect into the world. It is not just the case that things in time change, time itself undergoes a kind of 'change', whereby things that were future, become present and then recede into the past.

According to a B-theory of time, by contrast, events and times are ordered by a B-series ordering, but not an A-series ordering. The B-theory thus comes packaged with a rejection of the kind of dynamic picture of the universe embodied by the A-series. It is for this reason that the view is sometimes called the 'static' view of time. This name is somewhat unfortunate, however, as it gives the mistaken impression that the B-theorist rejects the existence of change. On the contrary, the B-theorist is fully committed to the existence of change. Change, for the B-theorist, consists in the possession of different properties at different times (sometimes called the 'at-at' notion of change). What the B-theorist denies, however, is that the kind of 'change' that characterises the dynamic aspect of the A-theory exists (hence their rejection of the existence of the A-series). Finally, according to a C-theory of time, events and times are ordered by a C-series ordering but not an A- or B-series ordering. The C-theory is therefore static (like the B-theory and unlike the A-theory) in at least some sense, but imbues time with less structure than the B-theory. Time, in both the A- and B-theories has a direction; not so according to the C-theory.

So much for the basics. Ultimately, the A-, B- and C-theories pertain to classes of theories: any theory that carries the above commitments can be classified as A-, B- or C-theoretic, or as an instance of one of these broad theoretical types. A great deal of work has gone into differentiating versions of the A-theory of time. A-theories of time differ with respect to their ontological commitments; in particular, with respect to the ontology of the past, present, and future. The most severe view is presentism.[7]

---

[7] For explication and defense of presentist versions of the A-theory see Deasy (2017), Ingram (2016), Pezet (2017), Paoletti (2016), Tallant (2012), Crisp (2003), Bourne (2006).

According to presentism (as we use the term—see Tallant and Ingram (forthcoming)), only present entities exist; past and future entities do not. But while only present entities exist, exactly which present entities exist is constantly in flux. As time passes, entities that were future become present, and as they do so they come into existence. As time continues to pass, entities that were present become past and go back out of existence.

Presentism can be contrasted with two other views. First, the growing block view. According to the growing block view, past and present entities exist, but future entities do not.[8] As time passes, the sum total of existing things grows, as more and more moments are added on to the existing 'block' of the universe. Second, the moving spotlight view. According to the moving spotlight view, past, present, and future entities all exist.[9] The dynamism encoded by the A-series is thus captured by a shift in which things are present. This, in turn, is to be understood in terms of a global shift in which existing things possess the properties of being past, being present and being future. As time passes, entities lose the property of being future, and then gain the property of being present. Soon, those same entities lose the property of being present and gain the property of being past.

In contrast to the various forms of the A-theory, variants on the B- and C-theories are more limited. The main form of the B-theory holds that all events and times can be ordered by a B-series only, and that all of the events and times between the Big Bang and the Big Crunch exist equally. The B-theorist also maintains that there is a direction to time: events and times are ordered as running from the Big Bang to the Big Crunch, rather than vice versa. This broad picture is sometimes referred to as the block universe, in virtue of the way that reality forms a four-dimensional 'block' of space and time.

As we will understand the B-theory, then, one is a B-theorist if one thinks that time has a direction, but having that direction is not the product of time passing. Time's having a direction might be a primitive or fundamental matter (Maudlin (2007), Oaklander (2012), and Tegtmeier (1996; 2009; 2014; 2016)) or it might reduce to some other, directed, asymmetric relation like causation (Mellor 1998; Le Poidevin 1991), or it might reduce to an asymmetric

---

[8] For explication and defense of variants of the growing block view see Rosenkranz and Correia (2018), Forbes (2016), Forbes and Briggs (2017), Tooley (1997), Forrest (2006), Button (2007).
[9] For explication and defense of variants of the moving spotlight view see Cameron (2015), Skow (2015), Deasy (2015), Miller (2019).

distribution of certain properties through time (such as, for instance, the increase of entropy away from a boundary condition (Loewer 2012; Albert 2000).[10]

As we are conceiving of the view, the C-theorist roughly agrees with the B-theorist when it comes to ontology: they both accept that all events and times exist equally. However, the C-theorist denies that there is any fact of the matter regarding the direction of time: time does not run from the Big Bang to the Big Crunch. These two events are simply two ends of a series that has no in-built direction.[11]

The way we are thinking of C-relations borrows from the way that we take Price (1996) and Farr (2012) to be conceiving of them: as a metric ordering of events along something like a dimension, or as part of something like a manifold, but in which that dimension is undirected. Picture a block universe. Given how we are thinking of things, if the B-theory is true then the block has built into it (either fundamentally or derivatively) a sort of arrow that points from the past to the future. By contrast, if the C-theory as we conceive of it is true, then the block has no inbuilt arrow: there is a four-dimensional block, with events and distance relations between them, and no fact about the direction along any of those dimensions, and hence no fact about which direction is future, and which past.

Of the theories just cited, the A-theory is taken by many to be the 'intuitive', 'folk' or 'common sense' picture of time.[12] Even non-dynamists often concede that they incur the burden of explaining why dynamism gets *something* right about how we ordinarily think about time, given that time is not in fact dynamical.[13] Of the particular A-theories cited, presentism is most often taken to be the view of time that the person on the street believes (or, at least, would endorse, were they to think much about the matter). Some growing block theorists disagree with this diagnosis, maintaining that theirs

---

[10]   In this regard, we depart from a looser usage of 'B-theory', in which the expression is used more or less synonymously with 'block universe theory'. On this latter looser reading it is not part of the B-theory that time has a direction. Since we think the question of whether it is part of the folk concept of time that time has, or must have, a direction, we want to carefully cleave the B-theory from the C-theory.

[11]   This way of conceiving of the C-series, and in turn the C-theory, is likely not what McTaggart, who introduced the terminology, had in mind. In fact, McTaggart uses the expression 'C-series' in somewhat different ways in different texts.

[12]   See Callender (2008; 2017); Smith (1994), Craig (2000), and Schlesinger (1994); Putnam (1967).

[13]   For ways of discharging this burden see Ismael (2012); Callender (2017); Miller, Holcombe, and Latham (2018).

is the view that accords closest to common-sense views about the nature of time (Broad 1923).[14]

The idea that presentism or the growing block view is the intuitive model of time has, for better or worse, come to play a substantive role in the debate over the nature of time. For some philosophers, the notion that presentism, say, is an intuitive view is taken to play a motivating role.[15] We should take presentism seriously, some maintain, because it is a piece of common sense. The appeal to the intuitive nature of presentism is not just made by presentists. Philosophers who criticise presentism sometimes do so in virtue of the fact that presentism is the common sense take on time. Hilary Putnam (1967), for instance, argues against presentism on these grounds.

As we shall see, there is sufficient variance in how people conceptualise time to undermine *any* appeal to 'common sense' or 'intuition' in the metaphysics of time. The empirical work that we detail below thus comes with a strong methodological recommendation: appeals to common sense should be banned in the philosophy of time.

2.3. Sixteen Hypotheses

In what follows we articulate sixteen hypotheses. The first twelve of these are sorted into four clusters of three each. In each cluster, we have three hypotheses: actual; conditional; and unconditional. The actual hypothesis is the hypothesis that actual time has some particular feature, F. The conditional hypothesis is that some feature F, is necessary for there to be time if actually F obtains. The unconditional hypothesis is that there is some feature, F, such that time obtains in any world only if F obtains.[16] We will unpack each of these in more detail shortly.

The first four clusters of hypotheses each take a different view about F. The first of these is the dynamical cluster, according to which F is the presence of temporal passage. The second of the four is the presentist cluster, according to which F is the existence of only present things. The third is the B-series cluster,

---

[14] As we shall see a bit later on, neither presentists nor growing block theorists are right. There is no one folk concept, and so no view can claim to be *the* common sense or folk view of time.

[15] For arguments of this kind see Zimmerman, (2008) Smith (1994), Craig (2000), and Schlesinger (1994). Though for an argument to the contrary see Torrengo (2017a).

[16] The reader might notice that this pattern of actual, conditional, and unconditional mirrors certain aspects of two-dimensional semantics. This is no accident, and is something to which we return shortly.

according to which F is the presence of a B-series. The fourth is the C-series cluster, according to which F is the presence of a C-series.

As noted there are sixteen hypotheses in total. The last four hypotheses are stand-alone hypothesises, and we will outline those once we have discussed the first four clusters, beginning with the dynamical cluster.

As noted, there appears to be a tacit view among philosophers that there is a univocal folk concept of time. Indeed, several philosophers have offered accounts of this concept either explicitly or implicitly, in part, or in whole. Insofar as these accounts tell us that certain conditions are conditionally, or unconditionally, necessary for our concept of time to be satisfied, they provide a useful starting point for our investigation. And so, in the first instance, we will outline different kinds of hypotheses about what the folk concept of time might be that have either explicitly been offered by philosophers, or are implied by extant philosophical discussions.

The first type of hypothesis we will consider arises in the context of McTaggart's (1908) infamous argument against the reality of time. There is no shortage of different ways in which to present this argument. Here is ours.

(1) In order for time to be real[17] the A series must be real.
(2) The reality of the A series requires temporal passage (such that events are first objectively future, then present, and then past).
(3) Temporal passage is incoherent.[18]
 Therefore,
(4) Time is unreal.

Key for us are (1) and (2). We take it that (1) and (2) constitute a conceptual claim that, in order for our concept of time to be satisfied, our world must contain this kind of A-series passage. As it happens, McTaggart also takes the obtaining of an A-series to be *sufficient* for time's existence, but that is less important for present purposes.[19]

---

[17] Here, we use 'real' since this is McTaggart's term. In this context we take it to be synonymous with 'exist'.

[18] In short, because to say that events are first future and *then present, and then past*, is to say that events go through the sequence: future, future-present, future-past. Since no event can be future and present, or future and past, so the A-series isn't coherent. We realise that we're compressing a lot here. The reader interested in the details of McTaggart exegesis is advised to read McTaggart's original text, as well as Ingthorsson (2016).

[19] Note that McTaggart does not believe the same thing about the B-series. The existence of the B-series, on McTaggart's view, is only sufficient for the existence of time if it comes coupled with an A-series. As a matter of fact, he maintains that the B-series is in fact sufficient for the existence of time, but only because he takes the B-series to be grounded in the A-series (plus the C-series) and thus assumes that if the B-series exists, there must be an A-series to support it.

The concept of time that appears in McTaggart's argument is what we will call *unconditionally A-theoretic*. Which is to say, it is a concept on which it is unconditionally necessary for time that there is an A-series.[20] We will say that C is unconditionally necessary for a concept of time to be satisfied, just in case that concept is satisfied in all and only worlds in which C obtains. By contrast, C is *conditionally necessary* for a concept to be satisfied, just in case if, actually, C obtains, then that concept is satisfied in all and only worlds in which C obtains. If C does not actually obtain, then that concept is actually and counterfactually satisfied in the absence of C.

Of course, there are many temporal dynamists who think that it is essential to time that it flows, but who do not think that this is a *conceptual* necessity. It might simply be, as a matter of metaphysics, that the A-series is essential to time. It is often unclear exactly what view different A-theorists take on this matter. For instance, Smith (1993), Gale (1968), Ludlow (1999), and Schlesinger (1980) think that we must posit an A-series because we cannot reduce A-theoretic talk to B-theoretic talk plus indexicals. If we think that our concept of time is intimately connected to our ways of talking about time and our position in it, and if A-theoretic talk is not reducible to B-theoretic talk, then perhaps indeed we have a reason to suppose that it is conceptually necessary that time is dynamical. In what follows, however, we won't assume this link between A-theoretic talk and the concept of time, seeing as it likely rests on controversial views about language and mental content.

Now, McTaggart, and most of these other authors, do not attribute the particular concept of time that they are working with to the folk. Nonetheless, and as already noted, the idea that the 'everyday' concept of time is A-theoretic *in some sense* is a common enough view among philosophers.[21]

Let's distinguish three hypotheses in this vicinity, from weaker to stronger:

Actual Time Dynamical Hypothesis: According to the folk concept of time, actual time is dynamical: (i.e. it is characterised by an A-series).

The Conditional Dynamical Hypothesis: According to the folk concept of time, the presence of a dynamical A-series is conditionally necessary for time.[22]

---

[20] In a similar vein to McTaggart, Williams (1998; 2003) holds that it is conceptually necessary that time passes. He, at the very least, takes it to be part of his concept of time that the presence of an A-series is unconditionally necessary for time.

[21] See Baron, Cusbert, Farr, Kon, and Miller (2015) for discussion.

[22] Baron and Miller (2015a) and Cusbert and Miller (2018) flirt with a view like this.

> **The Unconditional Dynamical Hypothesis:** According to the folk concept of time, the presence of a dynamical A-series is unconditionally necessary for time.

Both the conditional and unconditional dynamical hypotheses entail the actual time dynamical hypothesis if we hold fixed the further assumption that the folk suppose there actually to be time. The converse, of course, is not the case: the folk might largely suppose that time is actually characterised by an A-series and yet not think that it is conditionally or unconditionally necessary for time that it be so.

We will say that a folk concept of time is A-theoretic only if either the conditional or unconditional dynamical hypothesis is true of that concept. By contrast, we will say that someone is an A-theorist, or is a dynamist, just in case their (possibly tacit) model or picture of our world is closest to a dynamical model of our world. So, someone is an A-theorist just in case they hold that time in the actual world is characterised by an A-series. The A-theorist then might, or might not, have an A-theoretic *concept* of time. She will count as having an A-theoretic concept of time only if she thinks that the presence of an A-series is necessary for there to be time, either conditionally or unconditionally.

What we say here holds, *mutatis mutandis,* for the B- and C-series too. The B-theorist is someone who holds that, actually, time is characterised by a B-series. One has a B-theoretic concept of time, however, only if one holds that the presence of a B-series is necessary (conditionally or unconditionally) for there to be time. *Mutatis mutandis* for the distinction between being a C-theorist, and having a C-theoretic concept of time.

So much for the dynamical cluster. We turn now to the second broad type of hypothesis regarding the nature of the folk concept of time: the presentist cluster. As already discussed, a number of philosophers maintain that the folk notion of time is not just A-theoretic but *presentist* in particular. Presentists believe in the objective, mind-independent reality of temporal passage. They add to this claim, however, a specific view about temporal ontology: that only present entities exist. Past and future entities do not. Although no specific concept is attributed to the folk in print we can infer a broadly presentist conception of time from the many appeals to common sense. One way to put the point is in terms of tense: reality, on this view, is fundamentally tensed. So, while it is not the case that dinosaurs exist, it is the case that they used to. Similarly, future events don't exist, but will come into existence as time passes. For present purposes, we will focus just on the (non)-existence of the past and the future coupled with the passage of time. Again, then, we can delineate three hypotheses of varying strengths:

Actual Time Presentist Hypothesis: according to the folk concept of time, actual time is presentist (i.e. only present things exist, and which things are present, changes).

The Conditional Presentist Hypothesis: according to the folk concept of time, presentism is conditionally necessary for time.

The Unconditional Presentist Hypothesis: according to the folk concept of time, presentism is unconditionally necessary for time.

We now come to the third kind of hypotheses about the folk concept of time: The B-series cluster.

Many philosophers think time is B-theoretic: they think that the presence of a B-series is both necessary and sufficient for time.[23] Some also think that there is no possible world containing an A-series, for roughly the reasons that McTaggart provided. In what follows we will say that a world is B-theoretic just in case it contains a B-series but no A-series.

Of course, the popularity of the B-theory among philosophers doesn't mean that philosophers attribute such a concept to the folk, but perhaps the folk have such a concept nonetheless. That seems particularly plausible in light of the fact that contemporary presentations of time, especially by way of representations of time travel, often present other times as being just like other places to which one can travel. The picture of time that these presentations appear to be working with, then, is likely to be B-theoretic. And because the folk don't find these presentations of time counterintuitive, or difficult to understand we might suppose that these presentations capture something that sits at the heart of a folk concept of time.

At the very least we can extract from philosophical disputes about the nature of time a cluster of hypotheses according to which the folk concept of time is in some good sense B-theoretic. We will say that time is B-theoretic just in case time is characterised by a B-series, and is *not* characterised by an A-series.[24]

Again, there are three hypotheses corresponding to the actual, conditional and unconditional formulations of a B-theoretic concept.

---

[23] Defenders of variants of the B-theory include Mellor (1998), Callender (2017), Maudlin (2007), Tegtmeier (2009), Oaklander (2012), Le Poidevin (1991).

[24] So, if time is A-theoretic, it is not B-theoretic, even if the presence of an A-series entails the presence of a B-series.

Actual Time B-Theory Hypothesis: according to the folk concept of time, actual time is B-theoretic (i.e. it is characterised by a B-series but not an A-series).

The Conditional B-series Hypothesis: according to the folk concept of time, the presence of a B-series is conditionally necessary for time.

The Unconditional B-series Hypothesis: according to the folk concept of time, the presence of a B-series is unconditionally necessary for time.[25]

That brings us to the fourth kind of hypothesis: the C-series cluster. By analogy with B-theoretic worlds, we will say that a world is C-theoretic just in case it contains a C-series, but neither a B-series nor an A-series. Then there are three specific hypotheses corresponding to actual, conditional and unconditional formulations.

Actual Time C-Theory Hypothesis: according to the folk concept of time, actual time is C-theoretic (i.e. it is characterised by a C-series but not an A-series or a B-series).

The Conditional C-series Hypothesis: according to the folk concept of time, the presence of a C-series is conditionally necessary for time.

The Unconditional C-series Hypothesis: according to the folk concept of time, the presence of a C-series is unconditionally necessary for time.

C-theorists themselves certainly think that our world is C-theoretic; but we doubt they attribute this view to the folk. By contrast, the unconditional C-series hypothesis looks somewhat attractive. On the face of it, the presence of a C-series is plausibly a sort of minimal condition required for there to be time. If there are worlds that contain both a C-series *and* a B-series, those worlds contain time because they will contain a C-series, and in such worlds time is, contingently, directed. Likewise, if there are worlds containing a C-series and an A-series those are worlds that contain time, which, contingently, flows. The hypothesis that the folk concept is a concept on which the C-series is conditionally necessary for time is much less intuitively plausible, but is nonetheless worth testing.

---

[25] If, as a matter of necessity there is a B-series when and only when there is an A-series, then if the conditional dynamical hypothesis is true, then so is the conditional B-series hypothesis, and *mutatis mutandis* for the unconditional B-series hypothesis. In our studies, however, we present scenarios in which there is a B-series without an A-series. Hence our studies allow us to see whether, at least as a conceptual matter, the dynamical (conditional or unconditional) hypotheses are true, or, instead, if the B-series (conditional or unconditional) hypotheses are true.

That brings us to the last four hypotheses, which are different from the first twelve insofar as they do not mention the familiar notions of an A-, B-, or C-series.

The first such hypothesis is that the folk concept of time is a functional concept. Baron and Miller (2015a; 2015b) suggest that the folk concept of time is a functionalist concept. In what follows we take a functionalist concept to be a concept that is satisfied at any world if there is something in that world that plays some particular functional role. So, consider the concept <bachelor>[26]. <Bachelor> is not a functionalist concept: instead, it's a concept that is satisfied at any world just in case there is something that meets particular criteria: being an unmarried man. It would be a functionalist concept if, for instance, it were satisfied in any world by whatever plays some functional role: say, the role of being an attractive marriage partner. The difference between <bachelor> as we know it, and <bachelor> the functionalist concept, is that while on the latter view anything whatsoever can be a bachelor as long it plays the role of being an attractive marriage partner, on the former (familiar) view something can play the role of being an attractive marriage partner and not be a bachelor (by being a woman), and something can be a bachelor even if it fails to play the role of being an attractive marriage partner (by being a monstrous, psychopathic unmarried man).

So, the claim that time is a functionalist concept is the claim that something is time, in any world, just in case it plays some functional role R, where it is up for grabs what role R is.

Baron and Miller suggest that rather than supposing there are certain application conditions for <time> that appeal to the presence of a metaphysical structure, instead the concept might be one on which time is just *whatever it is that underlies certain experiences, or phenomena, in our world*. In particular, they suggest that for the folk, time is the thing that is responsible for our world seeming like *this*, where it's seeming like this is, roughly, its seeming to contain events that are ordered, with distance relations between them; its seeming to contain objects that persist; its seeming to contain events that are causally connected, where events earlier in the ordering cause those later in the ordering; its seeming as though we remember events that are in the past, but not the future; its seeming that we deliberate about events that are in the future, but not the past. Let's call this *a temporal seeming*. Baron and Miller reason that if this is right, the folk concept is a functionalist one: time is just whatever it is that plays a certain functional role: roughly speaking, the role of

---

[26] Henceforth we use angle brackets to pick out concepts.

grounding a temporal seeming (2015a). In what follows we will talk about what our temporal seemings *track*, where we suppose that our temporal seemings track some phenomenon P just in case P grounds our having the temporal seemings that we do.

If it turns out that a demon is responsible for the temporal appearances, Baron and Miller think that the folk will deny that the demon is time, since demons are fundamentally unlike what we take our temporal phenomenology to be tracking. In light of this, they argue that a functionalist analysis of time must include some additional constraint, so that it is not the case that *whatever* is responsible for the temporal seemings is time, regardless of what thing is like. In particular, it should not be that time is *fundamentally unlike* what we suppose our temporal phenomenology to be tracking. In turn, they think that something that plays that role would be fundamentally different from what we take our phenomenology to be tracking if it were not the kind of thing that is necessary for the existence of causation, persistence, and change (2015(a): 2435).

One way to think about this proposal is as a kind of constrained functionalist account—or, perhaps better, a sort of mixed functionalism on which time is whatever plays the role of grounding our temporal seemings, as long as it has features necessary for the existence of causation, persistence, and change.

We think that a better way to think of the view, however, is as one on which there are really two functional roles at issue: grounding our temporal seemings and grounding causation, persistence, and change. On this way of conceiving of the view, time is whatever it is that plays *two* roles: the role of grounding our temporal seemings, and the role of grounding causation, persistence, and change. In our study, we simplify this latter functional role and focus only on causation and change.

So, the version of functionalism we test is what we call Dual Role Functionalism.

> **Dual Role Functionalism**: Time is whatever it is that plays *both* the role of grounding our temporal seemings and the role of grounding causation and change.

A much weaker version of functionalism is what we call Seeming Role Functionalism. It is the following view:

> **Seeming Role Functionalism**: Time is whatever it is that plays the role of grounding our temporal seemings.

Recall, Baron and Miller think that Seeming Role Functionalism is false because they predict people will not conclude that if a demon plays the role of grounding our temporal seemings, then the demon is time.

We empirically investigate Dual Role Functionalism, though some of our results also speak to Seeming Role Functionalism. We can state the relevant functionalist hypothesis as follows:

The Dual Functionalist Hypothesis: According to the folk concept of time, time is whatever it is that plays *both* the role of grounding our temporal seemings and the role of grounding causation and change.

Are there functionalist conceptions of time that go beyond the ones discussed here? Possibly. One might, for instance, follow Callender (2017) in taking time to be the 'great informer'; the direction in which our laws take their simplest and most powerful form. Callender's account is supposed to apply to a scientific conception of time, but perhaps something like it could be turned into a hypothesis about the folk notion of time. Alternatively, there might be some other functionalist picture, not yet imagined by philosophers. We certainly can't rule this out (a point we return to briefly at the end of the chapter).

This brings us to our next hypothesis about the folk concept. Recently, Tallant (2018) has offered an account on which our concept of time goes unsatisfied just in case there are no present-tensed truths.

Tallant has (for these purposes) a fairly minimal notion of a present-tensed truth. There will be present-tensed truths if there are fundamentally tensed facts, and some of these are present tensed. But we take it that there will also be present tensed truths if it is, at some time, presently the case that certain truths obtain. We take this second notion to be compatible with a B- or C-theoretic view, on which present-tensed truths are simply those truths that obtain at some particular time. In sum: if a scenario has enough metaphysical structure to make it true of any x <x is present> then Tallant bets that the folk will judge time to be real in that scenario.

What we call the Actual Time Presence Hypothesis is the hypothesis that according to the folk concept of time, actual time is characterized by present-tensed truths. We didn't test this hypothesis because it seems most unlikely that the folk would deny that there are present-tensed truths, on the assumption that present tensed truths are not irreducibly present tensed, and so there are present-tensed truths even in B-theoretic and C-theoretic worlds. Second, we didn't test a Conditional Presence Hypothesis, according to which the presence of present-tensed truths is conditionally necessary for time in the folk sense. Tallant's account clearly states that the presence of such truths is necessary for time regardless of whether or not they actually obtain. Hence, we have just one Tallant-inspired hypothesis to test:

> **The Unconditional Presence Hypothesis:** According to the folk concept of time, present-tensed truths are unconditionally necessary for time.

So far, we have focused on specific hypotheses regarding certain aspects of the content of the folk concept of time, assuming that there is just one. As already noted, since the question of whether there is a single folk concept of time is one of the central empirical claims we aim to check, it might seem odd to spend so much time formulating hypotheses about the content of *the* folk concept, since this seems to just assume that there is a single folk concept in the first place. We should emphasise, however, that while we are interested in testing the above hypotheses, we do so in a way that enables us to simultaneously test two further hypotheses. These are the last two hypotheses of our group of sixteen. First, the *single concept* hypothesis:

> **The Single Concept Hypothesis:** the folk employ a single concept of time.

Full details about how this hypothesis is tested in parallel to the specific hypotheses discussed above can be found in Chapter Four, but the basic idea is simple enough. Our methodology for testing each of the individual hypotheses is to show participants different vignettes that describe the way the world might be actually, or how it might be counterfactually. Across the various studies we conduct, we test each of the fourteen hypotheses above (excluding the single concept hypothesis), and in doing so we gain information about the degree to which the folk endorse the various hypotheses we test. Variation between individuals does not, in all cases, provide evidence that those individuals are employing a different concept of time. For instance, Freddie might endorse a presentist view of time (that is, the view that actually our world is presentist) and Annie might endorse a B-theoretic view of time (the view that actually, our world is B-theoretic) and yet Annie and Freddie might have the very same concept of time: for they might agree about the conditions under which there is time, while disagreeing about what the actual world is in fact like. Many of the hypotheses that we test, however, are ones in which if individuals respond differently, this is good evidence that they have different concepts of time, for this is evidence that they disagree about the conditions under which their concept is satisfied. In short, the pattern of responses across the various studies can be used to confirm or disconfirm the single concept hypothesis.

Last, but not least, the *shared necessary condition hypothesis*:

The Shared Necessary Condition Hypothesis: all folk concepts of time share a necessary condition.

If there is just one folk concept of time then, obviously, the shared necessary condition hypothesis is true. If, however, there are multiple folk concepts of time, then it is an open question as to whether those concepts all overlap in terms of some shared necessary condition. Again, we use the pattern of responses across the Sydney Time Studies as the basis for confirming or refuting the shared necessary condition hypothesis. Part of our methodology is to show participants vignettes about how our world might be, or could have been, and ask them if there is time in that scenario. In the studies conducted, we do this while testing each of the hypotheses specified above (again, excluding the single concept hypothesis). What we should expect to see if there is some shared necessary condition, C, is widespread agreement that time does not exist, with respect to any vignette in which C is absent. If we find some such agreement across the studies, then that is evidence that the shared necessary condition hypothesis is true; if we don't find any such agreement then that is evidence against the hypothesis (of course, we shouldn't take the evidence to be conclusive, as there may be some feature of the folk concept(s) of time that we have failed to probe effectively).

We now have before us a number of distinct hypotheses about the folk concept of time. These hypotheses in no way capture all of the hypotheses associated with the views we have briefly touched on in this chapter. For instance, most A-theorists think that the presence of an A-series is not only necessary (conditionally or otherwise) for time, but also sufficient. Similarly, most B-theorists suppose the B-series to be sufficient for time, and most C-theorists think the C-series is sufficient for time. Likewise, Tallant (2018) argues that there is time if there are present-tensed truths and that such truths are not only unconditionally necessary for time, they are also sufficient.

As we noted earlier, however, we will predominantly focus on necessary conditions since it is these that aid us in determining the conditions under which our folk concept(s) go unsatisfied.

2.4. Empirical Results

To recap, we have sixteen hypotheses that we are aiming to test: fourteen hypotheses regarding the nature of the folk concept of time (assuming there is just one) and two hypotheses regarding variation in folk concepts of time (or lack thereof). We will start with the fourteen hypotheses about the folk concept. We will say that a hypothesis about the folk concept is supported when a statistically significant number of participants were shown to endorse a particular concept of time. In general, what we found is a dearth of support for any of the hypotheses identified above, with some notable exceptions (see Table 2.1).

The notable exceptions are the actual time dynamical hypothesis (top left), the unconditional C-theory hypothesis (fourth from the top in the last column) and Seeming Role Functionalism (second from the bottom in the second last column).

Before we discuss these two exceptions, it is important to clarify the results. First, it is important to say a bit more about Study 1, which effectively wipes out every hypothesis in the first column, bar one.

In Study 1 a range of different models of time were presented to participants in vignette form. The goal was to gauge the extent to which participants agreed that time in the actual world was accurately described by those vignettes. Participants saw vignettes representing a range of different dynamic—i.e.

Table 2.1. Support for Fourteen Hypotheses Regarding the Folk Concept of Time.

	Actual	Conditional	Unconditional
Dynamic Hypotheses	Supported (Study 1)	Not Supported (Study 2)	Not Supported (Study 2)
Presentist Hypotheses	Not Supported (Study 1)	Not Supported (Study 3)	Not Supported (Study 3)
B-Theory Hypotheses	Not Supported (Study 1)	Not Supported (Study 4)	Not Supported (Study 4)
C-Theory Hypotheses	Not Supported (Study 1)	Not Supported (Study 5 & 6)	Supported (Study 5 & 6)
Dual Role Functionalism		Not Supported (Study 7)	
Seeming Role Functionalism		Some Support (Study 7)	
Unconditional Presence			Not Supported (Study 8)

A-theoretic—theories (presentism, the growing block, the moving spotlight) and a range of different static—i.e. B-theoretic or C-theoretic—theories. Study 1 found a moderate convergence on the dynamic picture (about two-thirds of participants view time as dynamic, versus one-third who view it to be static) and then very little convergence beyond that: participants were roughly split across moving spotlight, growing block, and presentist accounts.

The convergence on dynamic pictures is important but not to be overstated: at best, it is convergence about time in the actual world and, moreover, it is still only a two-thirds majority. A large number of participants don't think that our world is dynamic. With respect to those who do think our world is dynamic, the divergence between various versions of the dynamic account is striking: there is very little agreement beyond a sense of time being dynamic that it is dynamic in any particular way. This is one of the pieces of evidence that we think cuts against any appeal to folk intuitions in support of particular A-theories (as foreshadowed in Section 3.1).

The substantial variation both in how participants see time actually and in what they think it would take for there to be time, provide evidence against the single concept hypothesis. These studies show that even across a relatively homogeneous population there is a wide degree of variability in people's judgments about the conditions under which there is time. Conceiving of concepts as we do, then, this is strong evidence of conceptual variability within the population tested.

Notably, in some studies participants disagreed about whether time exists under various discoveries that we could make about how things actually are, and then, further, disagreed about whether there is time in various counter-factual scenarios conditional on things actually being that way. Indeed, what we find is that some, relatively small, percentage of people think that the presence of an A-series is unconditionally necessary for time; some, also quite small, percentage of people think that the presence of a B-series is uncondi-tionally necessary for time. And a much larger proportion think that the presence of a C-series is unconditionally necessary for time. Jointly, this evidence suggests that there is no single shared folk concept of time.

Given that there is substantial variability in the folk concept of time, we cannot identify the necessary conditions under which 'the' folk concept is satisfied by simply unpacking that concept; there is no single concept to be unpacked.

Given that there is no single, shared folk concept of time, is there nonethe-less some necessary condition that is common to all folk conceptions of time? Based on the evidence we have gathered, the answer appears to be 'yes'. In

Studies 5 and 6, we found agreement among participants that a C-series is unconditionally necessary for the existence of time. This was so regardless of the more general disagreement participants displayed.

Actually, it pays to be careful here. The vignettes in question very clearly describe a scenario in which there is no C-series, and most participants then respond that there is no time in those scenarios. Nevertheless, one might wonder whether it's really the C-series itself that is necessary. The vignettes we offer participants do not contain a C-series, but they also leave out a lot of other structure. So, our experiments in this regard do not distinguish between two hypotheses: that our folk concept is one on which the C-series is unconditionally necessary for time to exist, and that our folk concept is one on which some other thing that is missing from these vignettes is unconditionally necessary for time to exist. There's a good reason why it would be difficult to empirically determine which of these hypotheses is the correct one. The trouble is that it is unclear how to provide a description that effectively isolates the C-series, by showing that only the C-series is missing (and nothing else). Given this, the evidence in favour of there being some shared feature to our folk concepts of time rather underdetermines what that feature is.

Finally, Study 7 found some support for a very weak kind of functionalism about time: Seeming Role Functionalism. This is a radical view indeed, on which time is just *whatever* it is that grounds our having the temporal phenomenology we do, *absolutely regardless of what that thing is like*. Our results provide some support for this version of functionalism, insofar as they suggest that people are insensitive to the features of what it is that their temporal phenomenology is tracking. Having said that, the results from our other studies clearly show that people are inclined to judge that there is no time present when a world lacks certain features (such as a C-series, or, indeed, in some case a B-series or an A-series). These results, jointly, tend to suggest that people have a fairly minimal concept of time, but that it is probably not as minimal as Seeming Role Functionalism predicts.

As noted, it could be that the folk notion of time obeys some other functionalist conception, one that is different from Seeming Role Functionalism and Dual Role Functionalism. Whatever that functionalist conception might be, however, it would seem that it too fails to be satisfied in certain situations. So even if the folk notion is some functionalist picture weaker than Dual Role Functionalism, there are still situations under which people would deny that time exists.

2.5. Immunity to Error

Now, what do the studies tell us about the immunity to error of our folk concepts of time? In pretty much every study we considered, there are individuals who are willing to deny that time exists, given certain discoveries about how the world is actually. Unsurprisingly, the nature of the discoveries lines up with the concepts at issue. At least some participants deny that time exists upon discovering that there's no A-series and only a B-series. Similarly, some participants deny that time exists upon discovering that there's no B-series and only a C-series. And some participants deny that time exists upon discovering that there's no C-series.

Now, in most of the studies the number of people willing to deny that time exists is quite small. Things shift substantially once we get to the studies involving the C-series. Here it turns out that there is a substantial number of people willing to say that time does not exist. Indeed, most A-theorists presented with a vignette that lacks a C-series are willing to flatly deny that time exists in that scenario. Even C-theorists are willing to deny that time exists if there is not even a C-series. This is different to the A- and B-series. While it is the case that some A-theorists will deny that time exists if there is no A-series, many continue to believe it exists even if only the B-series exists. The loss of the C-series, by contrast, seems to have a much more profound effect on individuals' judgements.

Note that this last point is subject to similar qualifications as those offered with respect to a shared concept of time, above. The vignettes used to represent the lack of a C-series also lacked other features. So, it might be that folk are having a strong response to the lack of a C-series, but it could also be that they are responding to the lack of some other feature from the relevant vignettes. The empirical work conducted cannot differentiate between these two options (an issue that we will return to later on, in Chapter Five). Still, whether it is the C-series or something else, it is clear that many (and in some cases, most) individuals will be willing to deny that time actually exists when presented with certain scenarios.

Given this, we are now in a position to offer three morals from the Sydney Time Studies. First, none of the folk concepts of time are completely immune to error. There is substantial evidence that individuals will deny that time exists, given the right prompt. Second, the loss of either a C-series or something nearby (that is also absent from the relevant vignettes) is a discovery that would likely lead to widespread abandonment of the folk concept of time. Third, this is true not just for C-theorists, who take the C-theory to be unconditionally necessary for time (and so for whom we would expect such a response) but for those who take a dynamical picture of time to be actual, who make up a two-thirds majority of the population. We expect that those who take there to be an actual B-series will hold a similar view, though this was not tested in the studies.

We submit, then, that there are discoveries that we could make about our world that would lead people to conclude that there is actually no time. Now, one might worry that there will be a divergence between what people say they would do in a series of studies, and what they would in fact do were certain discoveries to be made about the world. So, for instance, while many people *say* that they would deny that time exists upon discovering that there's no C-series, that's not in fact what they would do. Instead, they would continue to say that time exists, and simply take the new discovery in their stride. Because we can expect this kind of response, so the thought goes, the folk concept of time is immune to error after all.

We can see roughly two ways for this to pan out. First, it could be that the folk have a very minimal concept of time, along the lines of Seeming Role Functionalism (or some other functionalist conception).[27] And so it may then be that people's concepts of time are *so minimal* that they will say that time exists so long as the world continues to seem to them the way that it already does. That is, so long as there is no substantial shift in the way the world is experienced to be, people will continue to say that there's time no matter what discoveries they make about the world.

Second, it might be that the folk, in fact, have a concept of time that gives rise to the kinds of judgements we see in the studies, but that they will be willing to modify that concept on the fly in the face of new discoveries to preserve their belief in the existence of time. The idea is not that they will start employing a new concept. The thought, rather, is that there can be a kind of continuity of concepts through alterations to the concept in question. We might think of some scientific concepts as being like this, like the concept of 'contact'. Our concept of contact, one might argue, has survived through scientific revolutions that appear to render the concept unsatisfied. Thus, we continue to believe that things are in contact despite the fact that our folk concept of contact was, at one stage, radically at odds with the science, which roughly told us that there is no contact in the relevant sense. The folk concept of time might be like this: it may be that we are so invested in the concept, that we'll keep it alive no matter what by modifying parts of it in an ad hoc way.

---

[27] Note that we did find some support for seeming role functionalism in the population.

The concept ends up being something like Frankenstein's monster as a result, but no matter. The continued updating of the concept allows one to keep believing that time exists.

The folk might in fact behave in way that is contrary to the manner that our studies predict. We can't definitively rule it out. It may be that the folk in fact have a conception of time that is so flexible that it can survive the loss of the C-series. Thus, there may be some very weak functionalist conception that continues to apply even when there is no C-series, contrary to the evidence that we have gathered here.

It could therefore be that the folk will continue to say that time exists no matter what, and that the concept is in fact immune to error for reasons that are factored out by experimental conditions. But we shouldn't lose sight of the nature of this suggestion: it is an empirical bet about what would happen, were we to learn certain things about the world. It is an empirical bet that we have no evidence for, and some evidence against (in the studies discussed here). We are thus willing to offer a bet of our own in the opposite direction. Of course, we could be wrong. The only way to really know is to live through a discovery of the kind we are imagining and see what happens.

We close, here, by noting that this is an area for much further work. The history of the philosophy of time in the twentieth and twenty-first centuries has been influenced by a series of philosophical views about what the folk think about time: what it takes for time to exist; which view of time is most intuitive, and so on. What the data reveals is that philosophers' efforts to correctly divine the views of the folk a priori are, to put it gently, somewhat lacking. Quite independently of what one might wish to say about whether our folk concepts are in error, the folk view of time warrants further investigation.

3

Study Methodology

3.1. Introduction

In the previous chapter, we outlined the empirical upshot of a series of studies conducted on the folk concept of time. In this chapter and the next we present the details of these studies for the interested reader. This chapter focuses on background to the studies, along with broad methodological issues that frame their development. In the next chapter, we outline the studies themselves and take a closer look at their results. There is still much to say about these studies. Further information can be found in the published versions of these studies, and preprints where appropriate.

We begin by reminding the reader of some existing empirical work on folk concepts of time and explain why we must look beyond it (§3.2). We go on to outline some of the methodological difficulties that arise in any attempt to investigate the folk concept of time (§3.3) and explain how these difficulties may be overcome. We finish (§3.4) by outlining a broad empirical strategy for investigating the folk concept of time.

3.2. Existing Empirical Work

Why do we need to consider new, specifically philosophical, studies? Why not just explore findings from the pre-existing empirical literature about how people think about and represent time and use that to study the folk concept of time?

These are fair questions. There is indeed a lot of social scientific, neuroscientific, and psychological, research into a host of aspects of the ways in which we interact, (or appear to interact), with time.

At the psychological and neuropsychological end of things there is empirical work on how we process temporal information—this includes research on how we bind together information from different sensory signals to determine when that information is coming from the same place and time, and research into how we determine the order in which events occur, and the duration

Out of Time: A Philosophical Study of Timelessness. Sam Baron, Kristie Miller, and Jonathan Tallant, Oxford University Press.
© Sam Baron, Kristie Miller, and Jonathan Tallant 2022. DOI: 10.1093/oso/9780192864888.003.0003

between them (for an overview see Nobre and Coull 2010 and Holcombe (2015)) including some fascinating research on temporal illusions (see Eagleman 2010, 2011). For instance, it has been shown that the order in which events are judged to have occurred can be reversed, and that simultaneity judgements can be manipulated, as can judgements about the duration between events. It has also been shown that subjects can come to believe, falsely, that the effect of an action occurs *before* that action, even when the subject performs the action in question (Stetson et al (2006)).

At the philosophical end, there has been considerable empirically informed work on the extent to which the temporal structure of our experiences themselves are simply given by the temporal structure of the things of which they are experiences (see Watzl 2013; Lee (2014) Phillips (2014); Phillips (2010); Phillips (2011); Dainton (2000)). For instance, some aspects of this research have focused on trying to explain why we experience some events as occurring in succession (i.e. in some order) and why we experience some *as successions* (think here, of the experience of a sequence of notes *as a* melody, which is often thought to be an example of an experience *of a* succession, cf. Dainton 2000; 2008; Gallagher 2003; Le Poidevin 2007).

Importantly, however, for our purposes, the kind of research just outlined presupposes that there is temporal structure in the world. It asks how we come to know about that temporal structure, and then asks to what extent the content of our temporal experiences matches the temporal structure of the world. This research, which focuses on our temporal experience and whether or not it matches the external temporal structure of reality, is of little help to us in determining anything about the content of the folk concept(s) of time.

Philosophers have also appealed to empirical research in order to explain aspects of our temporal phenomenology other than order, duration and succession. In particular, many have thought that we experience time as dynamical—as passing—and that this is either because time does in fact pass or because we are subject to a pervasive phenomenal illusion (because time does not pass).[1] Among those in the latter camp there has been a sustained research project that appeals to the ways we process motion and change, and, in turn, to certain illusions related to these phenomena (such as the motion after-effect and phi motion (Nishida & Johnston, 1999; (Tyler, 1973; Steinman,

---

[1] Here, by temporal passage we mean what some mean by robust temporal passage: on this understanding, there is temporal passage just in case there is some objective fact of the matter as to which moment is present, and which moment that is, changes. Any dynamical theory of time—or, as we will also sometimes say, any A-theoretic view of time—is one on which time passes. We use 'dynamical theory' and 'A-theory' interchangeably in this chapter.

Pizlo, & Pizlo, 2000)) in order to explain why it seems to us as though time passes, even though it does not (Le Poidevin 2007, p. 76); Paul 2010, p. 346; Davies 1995, Schuster 1986, Skow 2011; Callender 2008; Hohwy, Paton, and Palmer 2016). Or, more recently, there are those who think that it does not, in fact, seem to us as though time passes, although we are inclined to believe that it *does* seem that way.[2] Proponents of this view have appealed to rather different empirical research to provide an alternative account of how we can be so wrong about the way things seem to us, temporally speaking (Hoerl, 2014; Torrengo, 2017b; Braddon-Mitchell, 2013; Deng 2013; Bardon 2013 p 95; Baron, Cusbert, Farr, Kon, and Miller 2015; Miller, Holcombe, and Latham 2018).

Again, though, this research is of little help to us. For while there are those who suppose that there is no temporal passage, and hence seek to explain aspects of our temporal phenomenology without appealing to passage, *everyone* in this debate assumes that there is some temporal structure, and no-one asks about the folk concept of time. As rich as it is, then, none of this work on temporal phenomenology speaks to fathoming any aspect of the folk concept of time.

So, what of research that focuses on the ways in which people represent time through spoken language, gesture, and diagram? From this research, we know that people tend to represent time spatially, or at least, to generate spatial representations when thinking about time (Gevers, Reynvoet, and Fias (2003); Fuhrman and Boroditsky (2010)), and we know that these representations vary cross-culturally in certain ways. We also know that the pattern of spatial metaphors that people use to talk about time varies across cultures (Núñez and Sweetser (2006); Borodistky (2001); Lai and Boroditsky (2013). In English (and other languages) spatial metaphors put the past behind the observer, whereas in Aymara, the future is behind the observer (Núñez and Sweetser (2006)). The same pattern is repeated when it comes to using gestures to represent the past and future, with the Aymara pointing in front of them to gesture to the past, and behind them to gesture to the future, the reverse of the pattern found among, say, English speakers.

We also know that cultural artefacts that represent time vary across cultures (Tversky, Kugelmass, and Winter (1991)). For instance, people who read text arranged from left to right, tend to pictorially represent the time-line as running from left to right (i.e. items to the left are represented as earlier than those to the right); those who write from right to left tend to represent

---

[2] For empirical research on the issue of whether it does indeed seem to us as though time passes, see Latham, Miller and Norton (forthcoming).

the time-line as running from right to left (Furhrman and Boroditsky 2010). Those who use a vertical axis when writing are more likely to represent the time-line time vertically (Boroditsky 2001; 2011).[3]

This work is important in explicating certain aspects of people's representation of time. It typically targets the ways in which people represent the time-line to others, (and presumably themselves), through language, picture, and gesture. This research allows us to answer questions about whether people represent the future as ahead of them or behind, above them or below them; to answer questions about whether people represent time as being linear, or circular, or something else, and, most notably, to answer questions about the extent to which these representations change cross-culturally.

It is, however, unclear just what connection these representations bear to an underlying folk *concept* of time. It might be, for instance, that there is a shared folk concept of time, but that cross-cultural differences result in relatively superficial differences in the way that people talk about, and represent pictorially and in gesture, the temporal axis. Alternatively, it might be that these differences are the result of there being quite different folk concepts of time at play. It certainly does not tell us what is *necessary* for time to exist. After all, that question is a modal one, and so cannot possibly be answered even by determining what people think time is actually like, in our world, and it is not clear that this research really even determines that. That is why we need additional empirical research to speak to this issue.

Finally, there is some recent research on folk temporal phenomenology, and, in particular, whether the folk experience time (or at least report doing so) as being dynamical or not. Several interesting studies out of Warwick (Shardlow, Lee, Hoerl, McCormack, Burns, and Fernandes (2020)) and the University of Sydney (Latham, Miller, and Norton (2020c)) aim to probe whether it seems to us, phenomenologically, as though time robustly passes. While this research comes closer to investigating the sorts of issues with which this book is interested, it still doesn't do the job required. For all of its merits, this work doesn't tell us anything about the folk *concept* of time. After all, someone could report that it seems to them as though time robustly passes, and yet that person might neither be an A-theorist nor have an A-theoretic concept of time. That is because even if it does seem to someone as though

---

[3] Interestingly, this has shown to be highly malleable: experimentally manipulating reading direction appears to change the way people spatially represent the time-line from past to future. Dutch speakers can reliably be brought to respond in a manner consistent with them spatially representing the past-to-future as running from right to left (Casasanto & Bottini, 2014) rather than from left to right, as they would without experimental manipulation.

time robustly passes, it doesn't follow that they *believe* time does in fact pass (and hence are an A-theorist) and still less does it follow that they hold that the presence of an A-series is necessary (conditionally or otherwise) for there to be time. So, while this work is illuminating, it does not illuminate the content of the folk concept of time. That is why we ran the studies we did, which focus on the concept of time.

We now need to address some worries that naturally arise from any empirical investigation into the content of our concepts, and, indeed, some that arise in particular with regard to the folk concept(s) of time.

3.3. Methodological Challenges

As in any empirical research, there are plenty of methodological challenges. Some of these challenges concern how to probe the content of people's concept of *time* in particular; some are worries about accessing the content of *any* concept, be it time or otherwise, and some are much more general worries about experimental work of this kind, whatever its aims. The replication crisis might be at the forefront of your mind, here, and you might be ill at ease with the prospect of social scientific research.

We cannot hope to defend experimental philosophy in full generality: indeed, we think that such a fully general defence would be a mistake. Some experimental philosophy is done well, and some is not. Some uses inappropriate statistical analyses; some is underpowered; some fails to get to the heart of the relevant philosophical question. Some, no doubt, will be replicable, and some will not. We do not wish to defend the bad; we aim to complete the good. What we can do is say something about more particular worries about these kinds of studies as we have used them, in the hope that this will go some way towards putting to bed certain concerns. So, each of the following sub-sections will articulate a worry about using experimental methods to uncover our concept of time, and then outline what we take to be the solution to that worry.

3.3.1. Accessing the Content of Concepts

There are many reasons one might be sceptical that empirical work of this sort can help us access the content of folk concepts. In what follows we will focus on four of these worries: that the concepts are too complex; that we have no access to the concepts; that we need but lack a robust analytic/synthetic distinction, and that the methods we deploy are simply too difficult.

To simplify matters a little, we will suppose that there is a tight connection between concepts, on the one hand, and linguistic items like terms and expressions, on the other. So, we will suppose that the meaning of some term like 'H_2O' just is the content of the concept H_2O. We allow that what 'H_2O' means in the mouth of any two individuals might be slightly different—they might speak slightly different idiolects—and hence their respective concepts of H_2O might be somewhat different. Further, we assume that facts about how people use a term (and the associated concept) play an important role in determining the meaning of the term, or the content of the concept. We will have much more to say about this shortly. For now, these assumptions are enough to raise problems for our approach.

3.3.2. Conceptual Analyses are Too Complex

Just think about the complex ways in which we use ordinary terms like 'table' or 'dog'. If we tried to set down a set of rules to codify this use, we might find it almost impossible. So, we shouldn't expect to be able to provide some nice set of necessary and sufficient conditions for something to fall under a concept as complicated as the folk concept(s) of time.

The fact that there is no neat set of such conditions doesn't mean that there is no conceptual analysis of a concept to be had. Perhaps such analyses are long and messy. Lots of things are long and messy. Some things are infinite: it doesn't follow that they don't exist. Then, by trading off some aspects of verisimilitude for simplicity we may be able to provide a useful *guide* to the content of the concept that is not long and messy.

But, for the sake of argument, let's suppose it is not possible to provide a set of necessary and sufficient conditions for something to fall under that concept or concepts (even an infinite set of conditions). Fortunately, this book does not aim to provide a conceptual analysis of the folk concept(s) of time. Instead, we are focusing on the much smaller job of discovering whether the folk concept of time is immune to error. Even if it's true that there is no way to produce a viable conceptual analysis of the folk concept(s) of time, it doesn't follow that we cannot show that there are discoveries that would lead individuals to deny that time exists.

3.3.3. No Access to Conceptual Content

A further worry concerns whether people are able to reveal the content of their concepts in experimental settings. There are two ways we might take this concern. The first is that people can't report the content of their concepts. This is simple for us to respond to; we do not ask people to report in this way—we ask them to *use* their concepts.[4] The second way to take this concern is that people are not in fact able to use their concepts in the way we require.

The concern here is that although people's general use of a concept is a good guide to at least some aspects of its content, we have little reason to think that people's judgement about scenarios described in vignettes in experimental setups are good evidence of their actual usage of a concept. Suppose, for instance, we present participants with a scenario in which it turns out that, actually, the wet potable stuff that comes out of taps and that is found in lakes, and so on, is discovered to be XYZ and not H_2O. Suppose we then ask them whether this would be the discovery that the wet stuff is not water or instead the discovery that water is XYZ.

One might agree that the way in which we would use 'water' in this circumstance is a good guide to certain aspects of the meaning of 'water' and the content of the concept of water. But one might be sceptical that what we *report* we would say, in that eventuality, is a good guide to what we would in *fact* say. For instance, suppose participants all respond that in that circumstance they would report that the wet stuff coming out of the taps is not water. How do we know that in fact, when faced with that circumstance they wouldn't continue to use 'water' to pick out that wet stuff?

To be sure, we know from work in behavioural economics that what people report their preferences would be, when given vignettes, and what people's preferences actually are, when given real world choices of the same kind, are closely matched (Johnson & Bickel 2002; Lagorio & Madden 2005). But we can't be *sure* that people can accurately simulate what they would say if certain things came to pass. At best, we have to take participants' responses as defeasible evidence for their use of a term. Still, all evidence is defeasible, and some evidence is better than no evidence at all. So, again, we will set aside

---

[4] We present people with descriptions, and we ask them whether their concept is satisfied by those descriptions. So, for instance, if we wanted to know about people's concept of table, we'd present them with a range of vignettes that describe arrangements of legs, flat surfaces, and so on, being used in various ways, and then we'd ask people whether the scenarios are ones in which there is a table present. People are readily able to use their concepts to determine whether a scenario is one in which there is a table present, even if they can't tell us the necessary and sufficient conditions for something's being a table (assuming there are such conditions).

this sort of worry. Philosophy rarely provides apodictic evidence for anything, and this book is no exception.

3.3.4. There's no Analytic/Synthetic Distinction

One might worry that the sort of methodology we are pursuing assumes a robust analytic/synthetic distinction. That's because we appear to be assuming that certain claims are analytic to the meaning of 'time', and hence are necessary for 'time' to have an extension, and for the concept of time to be satisfied.

If one thinks that there are no analyticities—for whatever reason—one will find this methodology suspect. For instance, one might follow Quine (1951) in thinking that meanings are holistic, and that nothing is truly analytic. Rather, some beliefs are more central than others and are less likely to be altered in the face of new discoveries or alterations to existing beliefs. Nevertheless, there is always *something* we could discover about our world, and some way we could choose to revise our beliefs, in light of that discovery which would result in those beliefs in the centre of our belief web being altered. In light of that, there are no analytically true claims: there are simply claims that it is very unlikely we will come to believe to be false. If so, there are no necessary conditions for a concept's being satisfied at a world; what might have appeared to be necessary for that concept to be satisfied, is not.[5]

---

[5] Let's suppose it's true that there is always *some* discovery we could make, and *some* way we could alter our web of beliefs, such that what appears to be analytic of some term turns out not to be.

Nevertheless, it might be that there is some condition, which, in all close possible worlds, given our web of beliefs, is such that if that condition were missing we would be inclined to judge that some concept is not satisfied. For instance, perhaps there are possible ways things could go in which it would turn out that '2+2=4' is false. If so, it's not necessary that 2+2 =4. Still, if the things we'd need to discover are things that are radically different from how we take things to be, or if the changes to our web of beliefs would be changes to something very central, then we can be confident that *most* of the ways things are likely to go, will be a way in which 2+2=4.

Now suppose we find that most of the ways things are likely to go are ways in which if condition C fails to obtain, most of us will judge that our folk concept(s) of time is not satisfied. Then it seems reasonable to conclude that very likely, if a physical theory is one in which C is absent, then that is a theory on which our folk concept(s) of time is not satisfied. Perhaps there is some, very unlikely, way things could go, and some very unlikely alteration we could make to the rest of our beliefs, that would result in us concluding that there is time if things are that way: but we have little reason to suppose that things would go that way.

If this is how one thinks of things, then one should interpret everything we say in this chapter and the next two as being not about a shared necessary condition, but about a shared condition, C, in which, under an array of ways things might plausibly go, it is very likely that most of the folk concept(s) are not satisfied if C is absent. It seems to us that this is of sufficient interest to warrant investigation. After all, what we want to know is whether, if certain scientific theories turn out to be true, our folk concept(s) of

However, we are disinclined to conclude that there are no necessary conditions for something to fall under any concept. We *do* think that what is necessary in order for many concepts to be satisfied depends on various facts about the actual world. We agree with Quine in at least this much: some conditions are not going to be *absolutely necessary*—necessary regardless of how things in fact are. In this, we follow a broadly two-dimensionalist picture of semantics like the ones developed in Jackson (2004), Chalmers (2004; 2012) and Chalmers and Jackson (2001). To see the idea, we will outline the Jackson-Chalmers picture, say something about how it connects to worries about necessity, and about how it connects to the methodology we deploy in our studies. This semantic picture is, of course, itself controversial. And so, we'll return to consider some more general worries one might have about it.

On a broadly two-dimensionalist semantic picture we distinguish two kinds of meanings of expressions: primary and secondary intensions. Very roughly, the secondary intension of an expression is what is often thought of as simply the intension of that expression: a function from worlds to truth-values (for sentences) and extensions (for terms). The secondary intension is, as it were, anchored by the way the actual world is. If, say, the actual world is such that the watery stuff is all XYZ, then the secondary intension of 'water' will (let's suppose) have XYZ as its extension in every possible world. If, actually, our world is one in which the watery stuff is H_2O, then the secondary intension of 'water' has its extension as H_2O in every possible world.

Imagine, then, a set of secondary intensions, each of which is anchored to a different actual world. To put it another way, think of all the ways the actual world might turn out to be for all you know (for all that you can rule out a *priori*) and now consider the secondary intension of 'water' on each of these different assumptions. One way the actual world could turn out to be is that the watery stuff is H_2O; another way it could turn out to be is that the watery stuff is XYZ (we've been massively misled so far, by science), and so on. Then we can ask ourselves, on each of these assumptions about the actual world, what is the extension of 'water' across the possible worlds?

We can say that we consider a scenario (or world) 'as actual;' when we are supposing that we have discovered that scenario to actually obtain, and we are considering a scenario (or world) as counterfactual, when we first fix the way the actual world is, and then consider that scenario as obtaining in a

---

time would be satisfied. That there might be very distant possible worlds in which a certain scientific theory turns out to be true, and in which we revise our beliefs in such a way that we are inclined to say that there is time, seems like something we might safely set aside for the time being.

non-actual world. Then the primary intension of a term is a function from each world considered as actual, to a truth-value or extension. So, for instance, suppose that whatever the chemical composition of the actual watery stuff in the lakes and coming out of the taps turns out to be, we are inclined to judge that that stuff is actually water. If we use the phrase 'the watery stuff' to pick out the stuff, whatever it is, that fills the lakes and comes out of the taps, then 'water is the watery stuff' turns out to have a necessary primary intension: it's true at every world considered as actual.

Further, suppose we are inclined to judge that *whatever* chemical composition that stuff actually has, all and only the stuff with *that* chemical composition in counterfactual worlds, is water. Now suppose, finally, that the actual stuff in the lakes and taps is XYZ. Then it will turn out that 'water' and 'XYZ' co-refer in every world. Indeed, it will turn out that 'water is XYZ' has a necessary secondary intension. It doesn't have a necessary primary intension though, since there are worlds considered as actual in which 'water is XYZ' is false (namely any worlds considered as actual, in which the watery stuff has a chemical composition other than XYZ).

One way we might express this is to say that the presence of XYZ is *conditionally necessary* for 'water' to have an extension at any world: it is necessary if actually, the watery stuff is XYZ, and otherwise it is not (indeed, otherwise, it might be impossible that XYZ is water).

On this semantic picture, we can see that there are plenty of things we could find out about the actual world that would change what we take to be necessary for a term to be satisfied. Conditional on the actual world being one way, a certain expression has one necessary secondary intension, and conditional on its being another way, it has a different necessary secondary intension. Thus, many expressions will not be absolutely necessary—necessary regardless of how things are actually—they will only be conditionally necessary. A natural thought, then, is that the content of any concept is the entire two-dimensional matrix, which tells us, for *every* way the actual world might be, for all we know, the secondary intension of the associated term. The entire two-dimensional matrix, that is, gives us both the primary intension of an expression, as well as all the secondary intensions that the term has, conditional on the actual world being thus and so.

So, to paint a full picture of our concept of water we need to know what would satisfy the concept given all of the ways the world could, for all we know, turn out to be: the way in which the watery stuff is XYZ, and way in which it is H_2O, and the way in which what is in the lakes is a different chemical composition from what falls from the skies, and so on. In turn, to investigate our concept of water we need to probe people's dispositions to use that concept across a range of hypothetical scenarios considered as both actual, and counterfactual.

This, indeed, is the methodology that we use in our studies. We present participants with vignettes that describe various scenarios that are relevant to time (more on this shortly). We begin by asking them which of the scenarios is most like our world. This tells us what participants in fact think our world is like.[6] Later, participants are re-presented with one of the vignettes and told that it describes the way our world has been *discovered to in fact be,* and are then asked whether or not there is time in our world *given this discovery.* This is to ask participants to consider a hypothetical scenario as actual. We then have two pieces of information: both what participants in fact think our world is like, and what they would say about our world given some supposition about how it is. That supposition may, or may not, correspond to how they in fact take our world to be. It's important to collect information about what participants in fact think our world is like, so that we can check to see how well they are simulating the world to be other than they suppose it to be. Finally, participants are re-presented with a vignette and told that it describes some world that is *not* actual—we tell them it is a parallel universe—and they are then asked whether there is time in that world. This is to consider the scenario as counterfactual.

This methodology allows us to gather data on how participants use a concept across a wide variety of discoveries that they can make about how things are actually, and, in turn, how those actual discoveries affect how participants use the concept when evaluating counterfactual worlds.

Notably, then, we don't assume that there are absolutely analytic claims about, in this case, time; we don't assume that there are claims that participants take to be analytic of time, utterly regardless of what the actual world turns out to be like. Rather, we allow that what turns out to be (secondarily) necessary to the concept of time being satisfied, might depend on what the actual world is like. This view, then, takes seriously the Quinean idea that analyticities are not, as it were, essentially analytic: they might only be analytic given that our world is thus and so, and so, perhaps, are not strictly analytic at all. Or, to put it another way, such views take seriously the idea that what is

---

[6] It's worth noting that participants don't need to have previously given any thought to the nature of time. It is enough that when presented with some descriptions of how our world might be, they can decide which of these is most like what they take our world to be, even if their taking it to be that way is entirely tacit.

necessary for a concept to be satisfied may not be necessary *simpliciter*, but might only be necessary given that the world is thus and so.

We, then, are inclined to talk about what is necessary for a concept to be satisfied, conditional on the actual world being thus and so, or what is *conditionally necessary* for that concept to be satisfied. To pursue the project of this book we do not require that we find some condition, C, (shared or otherwise) which must obtain for the folk concept(s) to be satisfied, regardless of how things are actually. Of course, there may be such a condition, and we intend to see if there is. If a condition is one which, according to a folk concept, must obtain in order for that concept to be satisfied, regardless of how the actual world is, (at least with regard to time-relevant features) then we will say that such a condition is *unconditionally necessary* for that folk concept of time. Our methodology is designed to allow us to probe the folk concept(s) for both unconditionally and conditionally necessary conditions. Which, if any, we find is an empirical matter. In what follows, then, we set aside concerns that no necessary (or sufficient) conditions can be specified for the satisfaction of our concepts, because there is no robust analytic/synthetic distinction and hence nothing is unconditionally necessary for the satisfaction of any concept.

3.3.5. That Methodology is Too Difficult

At this point two further concerns arise. We've just briefly described a methodology to probe whether a condition is conditionally or unconditionally necessary for some concept to be satisfied. But now one might worry that this very methodology places too great a cognitive burden on participants, and consequently can't be expected to yield reliable results. We ask participants to imagine some scenario, and then to suppose that this scenario has been discovered to be actual. We then ask them to use their concept to determine whether the scenario described contains time. We *then* present still another scenario, and ask them to imagine that it describes a non-actual (parallel) world, and ask them to use their concept to determine whether the non-actual scenario described contains time, conditional on the first scenario describing how things actually are. But why think participants are able to successfully complete either, let alone both, of these tasks? Moreover, why think that even if people can consider a scenario as actual, or as counterfactual, that they are any good at all at determining what they would in fact say were they to come to discover that scenario to be actual, or what they would say about some non-actual scenario, conditional on some other scenario being actual? Surely, we are asking participants to engage in imaginative tasks with which they are extremely unfamiliar. So the worry, here, is that although participants appear to be following the instructions we give them, in fact they are failing to do so properly.

Let's call people's capacity to take a world other than the one they suppose to be actual, as actual, and then evaluate claims both in that world, and in a counterfactual world, actual-world context shifting. The key worry, then, is that people are not able to actual-world context shift: they are always evaluating counterfactual worlds as though the actual world is as they in fact suppose it to be. Now, if in our studies participants' judgements about whether there was time in a counterfactual world varied depending on which world they were told is actual, then this would provide some evidence that people can actual-world context shift. For their doing so would clearly be the best explanation for that pattern of responses. In our studies, however, we find no such pattern of responses. So the data leave entirely open that our participants are unable to actual-world context shift. While we have concluded, on the basis of this data, that people's concept of time is not sensitive to the way the actual world is, perhaps this is a mistake, and the concept is sensitive in this way, it is just that people cannot perform the necessary cognitive gymnastics to show that this is so. In what follows we will offer a number of considerations, which, jointly, speak against this hypothesis.

First, stepping back from the more complex question of whether people are able to actual-world context shift, let's focus just on the question of whether people can even simulate scenarios as actual, or as counterfactual. Fortunately, there is already some research of this kind into concepts other than time, which suggests that people do have the ability (though perhaps imperfectly) to simulate scenarios considered as both actual and counterfactual. First, there is very general evidence arising from the empirical investigation of causal reasoning, which shows that people are able to represent a range of counterfactuals in the form of counterfactual interventions. That is, we represent that *if* we intervened on some part of the world (though we in fact do not) we would thereby intervene on some other part of the world (Kushnir, Gopnik, Lucas, & Schultz, 2010; Lagnado & Sloman, 2004). For instance, when people are asked what will happen if an experimenter intervenes in the world in a particular way, there is good evidence that participants imagine the intervention and its consequences (Sloman, 2005). So, we know that, quite generally, people are able to simulate a range of ways things are not, and think through what would happen, if things were that way.

Often, of course, counterfactual reasoning requires that we suppose the world in question to be just as we take the actual world to be, except for the fact that the relevant intervention is performed. Equally, though, such reasoning has been shown to play a role in our analyses and understanding of historical and political events, whereby we imagine, or simulate, how history would have been very different from the way it is, had some earlier event been different (Tetlock and Belkin 1996). This reflects a general capacity to imagine that things could have been very different indeed from how they in fact are. Moreover, we know that people are able to draw fairly fine distinctions between counterfactual scenarios. For instance, people's judgements are sensitive to the difference between counterfactual *interventions* and counterfactual *observations* (Sloman 2005:77–78).

In addition, we know that future planning is based in part on our ability to construct mental models of the different ways things might go (Johnson-Laird 1983). On at least one way of thinking about how we do so, these different mental simulations are simulations of the ways things might go, for all we know: that is, they are simulations of what we might discover about the actual world, rather than imagined counterfactuals. That is to say, these are scenarios considered as actual. There is some evidence that this is indeed the right way to think about such cases. While there is neurophysiological evidence that episodic counterfactual thinking, episodic memory, and episodic 'future' thinking share common neural substrates—because people with impairment in one area typically show impairments across all three (see Hooker, Roese, and Park, 2000)—there are noteworthy differences between episodic counterfactual thinking and episodic future thinking. For instance, episodic counterfactual thoughts are experienced with less emotional intensity than episodic future thoughts (De Brigard and Giovanello, 2012), and repeatedly simulating episodic counterfactual thinking tends to decrease its perceived plausibly, whereas repeating simulating episodic future thinking increases it (De Brigard, Szpunar, and Schacter, 2013). This is at least consistent with the idea that some of these imaginings and simulations are simulations of counterfactual scenarios, and some as actual scenarios, and that we treat these simulations differently (or at least, that they have somewhat different broad effects on human psychology).

A final piece of general evidence that is relevant here is that experimental work strongly suggests that people understand counterfactual conditionals by imagining two possibilities, the actual world and the counterfactual alternative (Thompson and Byrne, 2002; Santamaría, Espino, and Byrne, 2005; De Vega,

Urrutia, and Riffo, 2007). This suggests that counterfactual reasoning already involves holding fixed that a particular world (or scenario) is actual, and then imagining a counterfactual scenario that differs from the actual scenario in some way. This in turn suggests that we are able to keep in mind both a scenario considered as actual, *and* one considered as counterfactual, and indeed that our evaluation of counterfactuals is already tied to an assumption about which scenario is actual.

Of course, none of this shows that people are able properly to imagine that *actually*, things are other than they supposed, and *then* evaluate claims both at the actual scenario and at a counterfactual scenario. That is, none of this shows that people are able to actual-world context shift.

What would such evidence look like? Well, if people differently evaluate a counterfactual claim depending on which world they are *told* is actual, then it seems very likely that the explanation for this is that they are able to imagine each of the different scenarios as actual, and then, on the basis of this, evaluate the relevant counterfactual claim. So, evidence that people's judgements about a counterfactual claim are sensitive to which scenario they are told is actual, is powerful evidence that people can actual-world context shift. Fortunately, there is evidence of just this sort in the free will literature.

Within the literature on free will, experimentalists have been interested in whether or not non-philosophers have an incompatibilist concept of free will—that is, a concept that is not satisfied at any world that is deterministic—or a compatibilist one—a concept that is satisfied at (some) deterministic worlds.

Results of experimental work are initially rather puzzling. It seems as though there is excellent evidence in favour of non-philosophers' concept of free will being a compatibilist concept (e.g., Nahmias, Mossis, Nadelhoofer, and Turner, 2005; 2006), and also evidence in favour of it being an incompatibilist concept (e.g., Nichols and Knobe, 2007). For example, Nahmias et al. found that a majority of people judge that someone acts freely when presented with a scenario describing that person's action across numerous descriptions of our world being deterministic: the compatibilist response. Meanwhile, Nichols and Knobe found that a majority of people, when presented with two hypothetical scenarios, one deterministic and one indeterministic, judge that our world is most like the indeterministic scenario. Furthermore, when asked whether people in the deterministic scenario are morally responsible (a proxy for free will) the majority of people judge that they are not: the incompatibilist response.

Reflection on differences in experimental set up (Roskies and Nichols, 2008; Latham, 2019) identified a slight yet important difference which could explain the apparently conflicting results and show how they can be reconciled. While Nahmias et al. presented some of their scenarios as ones in which the actual world is described as being deterministic, Nichols and Knobe presented some of their scenarios as ones in which a hypothetical scenario is described as being deterministic. Roskies and Nichols found that people's free will judgments when assessing a deterministic scenario differed significantly as a function of whether that scenario was described as actual, or as hypothetical (i.e. counterfactual). Participants who were assigned to make free will judgments when the deterministic scenario being evaluated was actual, were significantly more likely to agree that the scenario contained free will, than were those who were asked to evaluate the scenario as counterfactual.

Thus, Latham (2019) presented people with an *indeterministic* scenario described as actual, and asked them to evaluate both that scenario, and a deterministic scenario described as counterfactual, and *vice versa*. Latham found that people's responses to deterministic scenarios differed significantly as a function of whether they were evaluating the scenario as counterfactual or as actual. Specifically, people responded that indeterminism is only necessary for free will if it is actual, but if the actual world is deterministic, then determinism is compatible with free will. These results strongly suggest that, at least with regard to these scenarios, people are able to imagine a scenario as actual (even if it is not the way they in fact suppose our world to be) and then to evaluate counterfactual scenarios conditional on that scenario being actual: people can actual-world context shift.

Having said that, there is room for caution here. We grant that it might be harder to consider ways the world might be actually and counterfactually when we are asking people to think about time. It might be, for instance, that imagining that we discover our world to be deterministic (when we think it indeterministic) is easier than imagining we discover our world to be non-dynamical when we think it dynamical, and *vice versa*. But in the absence of any evidence that this is so, we will suppose that people are able to follow the methodology we employ.

3.3.6. Conceptual Change, Content, and Some Examples

At this point it is worth clarifying our methodology by working through some examples. First, it's important to remember that (on our view) the content of concepts is something like a primary intension, or a referencing fixing description. This means that what people *believe* about what satisfies their concept can turn out to be mistaken. Consider, again, the case of colour. Let us suppose that it is part of a folk view of colour that colours are intrinsic properties of objects, spread evenly across their surface. Does this mean it is part of the concept of colour that colours are like this? On our view, this is only part of the *concept* of colour if we are inclined to judge that if actually there are no such intrinsic properties, then there are no colours. Otherwise, this is simply a belief about what colours are, in fact, like. In this case we take it to be very likely that this is not part of the concept of colour, since we think people are in fact disinclined to conclude that there are no colours if there are no such properties. That is not to say that it might not be a very surprising discovery. To find out that colours are very unlike what we thought, when we come to learn that there are no such intrinsic properties, may count as a significant surprise. But that we can make that particular surprising discovery depends on the fact that we still take our concept of colour to be satisfied, albeit satisfied by something different from what we thought.

Importantly, then, and as the colour case illustrates, in the case of time we can discover that time is quite different from what we thought, consistent with our concept of time still being satisfied.

Given this, a question arises: what differentiates a case where a concept is satisfied in a surprising way from a case in which a concept changes, and from a case in which a concept is unsatisfied?

We think there are three possibilities in the temporal case. First, suppose that participants judge that there is no time in a certain scenario, S, considered as actual. Now suppose that subsequently the scenario turns out to be actual and the same people judge that actually there is time. We think the best interpretation of this data is that all along these people had a concept of time that would be satisfied in the (actual) scenario S, and they were mistaken when they judged, in the experimental setting, that there would be no time in that scenario. In this case judgments about what they would say if the world turned out to be a certain way, were not good guides to how people did in fact judge things to be when the world turned to be that way. All along it was the case that temporal realism would be vindicated if our world turned out to be a world of the kind described by the scenario S.

Compare that case to the following one. Under experimental conditions people judge that there is no time in a certain scenario, S, considered as actual. Now suppose that subsequently the scenario turns out to be actual, and the same people continue to engage in temporal discourse, whilst explicitly judging that there *really is* no time. Later, these people come to judge that there is time. We think that in this case people really did have a folk concept that was not satisfied by the scenario S. That is why they made those judgements in the experimental set up, and why they were inclined to judge that there was no time when it turned out that our world accords with S. Still, the pressure to continue to use temporal talk resulted in them *changing* their concept to a more liberal one that was satisfied in the actual world. That is why later, they come to judge that there is time. This is a case in which the folk concept, as it stood, was not satisfied by the scenario S, but people eventually moved to a new concept.

Finally, there is a third option. Under experimental conditions people judge that there is no time in a certain scenario, S, considered as actual. Now suppose that subsequently the scenario turns out to be actual and the same people continue to engage in temporal discourse. These people explicitly judge that there really is no time, and that there really are no temporal relations. They go on making those explicit judgements, but yet they continue to use the same language they always did.

In this case we think the best interpretation is that these people all along had a concept of time that is not satisfied by the scenario S. In continuing to use the language they do, these people are engaging in a kind of temporal fictionalism, in which they act and speak *as though* there is time, whilst all the while their concept of time is unsatisfied (for discussion of temporal fictionalism see Baron, Miller and Tallant (2021).

Now, prior to the discovery in question people in all scenarios judged that their concept would not be satisfied were the relevant discovery to be actual. In two of the three cases this judgement was good evidence about their under-lying concept. It is just that in one such case people later went on to change their concept. In one case the evidence was misleading, and it was always the case that their concept would be satisfied in a given scenario.

Our key point is that these cases are different and we could come to have evidence one way or another about people's concepts. Of course, we cannot know, *now*, what people will actually do and say, if they come to make these kinds of discoveries. But assuming people's experimental judgments are good evidence about what they would do, we can come to know certain things about the concept in question. What we cannot know, at this stage, is whether if it turned out that people's concept did go unsatisfied they would be inclined to change that concept (say) or continue to use temporal talk despite taking their concept not to be satisfied. This is a limitation of the current project to be sure, but not, we think, a fatal one. We think that people's judgements are good enough evidence to proceed with, even if there is scope here for error.

3.3.7. Participants, Populations, and Problems

A very different concern one might have about empirical research of the kind described here concerns the quality of the participants: they might not *really* be chosen randomly; they might not be appropriately concentrating, or taking seriously the task; they might have very different background knowledge about science and philosophy, and they might, in fact, be online bots. These are all legitimate concerns, but all concerns that can be set to one side.

On randomness: In all the work we report, large groups of participants (US residents) are given online vignettes, and then asked a series of questions. These participants are recruited through MTurk: a large pool of people in the US who undertake online experimental work for payment. While people signed up to MTurk may not be a completely random sample of the population, they come from a very wide demographic. That demographic information is available in the complete methodologies and statistical analyses that can be found in the references we provide in Chapter Four.

On bots: in some recent online studies run through MTurk, online bots, rather than human participants, have been answering the questions. However, those studies were not ones that used premium participants. Further, and unsurprisingly, bots tend to fail attentional check questions, and hence overall receive quite low-quality ratings. Thus, the safeguards described below for selecting the right participants should do away with concerns about our data including responses from bots.

On attention: our participants were all 'premium' participants: this puts them in the top 5 per cent of participants enrolled in MTurk. This means that they have to have successfully completed many such assignments, and that a vast majority of the assignments they complete, must be completed successfully. This helps to ensure that in general these participants are likely to be attending to the task at hand, and faithfully following instructions. Our studies also include attentional check questions throughout. This is to avoid including data from participants who are not attending to the vignettes or the questions we ask. We also included a set of comprehension questions in all the studies. These questions aimed to determine whether participants understood the vignette they read. We then eliminated data from participants who did not comprehend the vignettes. So, in each study we started with a large pool of participants, and then eliminated participants who did not understand the vignettes with which they were presented, or who failed the attentional check question. We also checked to see if there was any correlation between the responses participants gave, and whether or not they comprehended the vignettes; this is to rule out its being the case that our remaining sample is not random because, say, more people who think that our world contains an A-series are eliminated for lack of comprehension. We found no such effect.

In all, then, while no online experimental work can be certain that participants are adequately paying attention, or that the sample is truly random, we set the bar high with regard to using data from premium participants.

On background knowledge: most of the experiments run through MTurk are in psychology or behavioural economics, so there is little reason to suppose participants have any prior familiarity with philosophy beyond what you might expect in a random selection of the community. In our first study, we also ask participants a range of questions about their familiarity with science, especially physics and science relevant to time. We then correlated the responses to these questions with participants' responses to the vignettes. We will return to this study in detail in Chapter Four. For now, it's worth noting that we found no correlation between people's level of background knowledge of science, and of the physics of time, and whether they understood the vignettes, or which vignette they thought described a scenario that was most like our world.

Importantly, we can only make claims about the population from which the sample was taken: US residents. We cannot generalise these results to other populations. It seems likely that, for example, UK and Australian residents will respond in a similar manner, but we don't know that; we certainly cannot make assumptions about other, quite different, populations. So insofar as we talk about there being a number of folk concepts in the population, we are only talking about the population we tested. It may be that there are other, distinct, folk concepts in other populations.

Here, then, is one important limitation of the studies we performed: since they can only tell us about the population sampled, we cannot be sure that views about whether time exists in certain scenarios will be widely shared *across* populations. This leaves open the possibility that populations outside the US will not deny that time exists if, say, they are presented with a certain kind of scenario. Thus, it does not follow that a majority of people will conclude that there is no time in certain cases. At best, we can say that a majority of the population tested will conclude that there is no time.

In an ideal world, it would be nice to be able to address the question of whether our results generalise to other populations. But clearly this is far from an ideal world, and one of the difficulties with empirical work is that it is always limited in various ways. All we can say is such is life, and remind the reader to bear in mind the scope of our conclusions in what follows.

3.3.8. Time Neutrality

There are additional challenges that we face when attempting to discern necessary (conditional or otherwise) conditions for a folk concept of time to be satisfied. Consider a non-temporal case. How would one go about testing people to see whether they will think their concept of water is satisfied in a counterfactual world in which there is no H_2O? Presumably, though very roughly, one describes a scenario—without using the term 'water' in the description—in which the watery stuff is of some chemical composition other than H_2O, and in which the scenario is described as non-actual, and one then asks participants whether in the scenario there is any water.

By parity, if we want to know whether the presence of feature F is necessary (conditionally or otherwise) for there to be time, then we will need to describe a scenario—without using the term 'time' or related terms—in which feature F is absent, but some other relevant features are present, and then ask participants whether the scenario described contains time. The challenge, then: describe a scenario as being one in which F is absent, and in which we do not mention time, or temporal relations, or related terms, in such a way that participants can understand what is being described, and can do so in a way that allows them to then determine whether or not they think that scenario is one in which there is time. This is a difficult task. We proceeded as follows.

In our first study, we present participants with six vignettes, each of which describes a scenario in which a standard metaphysical theory of time obtains. In that study, we present participants with a presentist scenario, a growing block scenario, a moving spotlight scenario, a B-theoretic scenario, a C-theoretic scenario (more on this shortly) and a quantum gravity scenario (more on this also). These vignettes describe each scenario using temporal locutions. They mention times, durations, relations of earlier-than and later-than, the present, the past, and the future. Our aim, in that study is to find out which of these scenarios participants think is most like the actual world.

These vignettes clearly cannot be used when our aim is to determine whether some feature is necessary (conditionally or otherwise) for time. For if we describe a world using temporal locutions, this is like describing a scenario in which the watery stuff is *described* as being water, and then asking participant whether there is water in the scenario.

So, in the remaining studies we used the same basic vignettes as in our first study, but we stripped out all the temporal language. Towards the end of each vignette we introduced locutions such as: 'some scientists, philosophers and theologians in universe[7] C/D/E think that…' and then the ellipses included a description of the scenario using some temporal language.[8] The aim of this addition is to allow participants to see that the features we have described in non-temporal language *might* be features that are in fact temporal, but at the same time leaving it open that participants might judge that these features are not temporal, and that the scientists in question have made a mistake. Then participants are asked to determine whether there is time in the scenario thus described. We call vignettes like this *time-neutral* vignettes, since they are written in such a way that they do not presuppose that there is time in the scenario described. Again, we present participants with comprehension questions about these vignettes in order to make sure that they understand what is explicitly said to be true according to the scenario described.

3.4. Overview of Empirical Approach

So far, we have responded to a range of concerns about how to go about probing the folk concept(s) of time. Hopefully what we have said so far has already gone some way towards articulating the sort of research program that we pursued.

Of course, we cannot hope to present people with *every* possible hypothetical scenario. So, for all we know, there may be conditions that are relevant to whether people's concept of time is satisfied which we do not experimentally probe. For instance, we do not experimentally wiggle whether there is vegemite present in any scenarios considered as actual or counterfactual. Yet for all we know, it might be that many people take their concept of time to be satisfied only in a scenario that contains vegemite, or only in a counterfactual scenario that contains vegemite, conditional on the actual world containing vegemite.[9] If vegemite were, for instance, conditionally necessary for time, then this is not something our studies would reveal. Since the scenarios we

---

[7] We don't talk of possible worlds in the studies, but instead, we talk of universes.

[8] So, for instance, the analogue in the case of water would be that we first describe a world in which the clear, drinkable, liquid that falls from the skies and fills the lakes, etc., is of chemical composition XYZ, and then later say that some scientists (etc.) in that universe think that the universe is one that contains water.

[9] We do not assume that all readers will know what vegemite is. UK readers can substitute 'marmite' for 'vegemite'. Other readers can imagine clearing the muck from the bottom of a vat used for brewing beer, and spreading it on toast.

present to participants are silent on the status of vegemite, it is likely that most participants will assume that any scenario that they are considering as actual is a scenario that contains vegemite. They will then evaluate any counterfactual scenarios, conditional on vegemite actually existing. Perhaps, though, they would evaluate those scenarios differently if we were to tell them that, actually, vegemite has been discovered not to exist. We choose this example for obvious reasons: on the one hand, it's extremely unlikely that vegemite is necessary (conditionally or otherwise) for time; on the other hand, it's clear that our studies wouldn't reveal that vegemite is at least conditionally necessary for time, even if it were. What's true of vegemite is true for a variety of other less bizarre conditions that we also don't test.

All we can tell on the basis of our studies is that, in the population tested, there are certain conditions which, when absent from a scenario (considered as actual or counterfactual), result in most people being inclined to say that the scenario does not contain time. Perhaps there are other conditions we have not tested, and were we to test them it would turn out that they, too, are necessary (conditionally or otherwise) for time. If there are such conditions, and if they are widely shared, then there is scope for our concept(s) of time to be in error on grounds other than those we present. We have no reason to suppose there are such conditions, but, again, the reader should bear this in mind in what follows.[10]

Returning to our broad methodology, then, it is worthwhile at this stage introducing some terminology that will be of use in the following chapter. We have already said that some condition might be *conditionally* necessary for time. More carefully, we will say that according to a particular folk concept of time, FC, a condition, C, is *conditionally necessary for time*, just in case FC is satisfied in all and only worlds in which there is some relevant metaphysical structure that has C, conditional on C actually obtaining, and not otherwise. Here, we talk of some relevant metaphysical structure having C, rather than C merely obtaining, since otherwise it will turn out that, for instance, '1+1=2' is unconditionally necessary for time, since participants will judge that there is time in all and only scenarios in which 1+1=2. And while this might be a trivial unconditionally necessary condition for time, it's clearly not the kind of condition in which we are interested. So, for 'relevant metaphysical structure'

---

[10]   In some respects it doesn't much matter if there are: we argue that the shared necessary condition we discover is one that is absent from our world if our world is as these scientific theories says. So, at most, a discovery of further conditions of this kind would over-determine its being the case that our folk concept(s) of time are in error.

just read something quite loose, like 'something that a reasonable person could think is relevant to the obtaining of time'.[11]

By contrast, we will say that according to a particular folk concept of time, FC, a condition, C, is *unconditionally necessary for time*, just in case FC is satisfied in all and only worlds in which there is some relevant metaphysical structure that has C, regardless of whether C actually obtains. So, for instance, consider a theory of time that posits an A-series. This is the view that time passes in a robust manner: its passage consists in some change in which moment is present.[12] Then it might be that the presence of an A-series is conditionally necessary for time: perhaps each of our folk concepts of time is such that we judge there to be time in any world only if it contains an A-series, but we judge this only if actually there is an A-series. If our world contains, say, a B-series but no A-series, then we do not judge that there is actually no time: instead, we conclude that the A-series is not necessary for time. Alternatively, perhaps the presence of an A-series is unconditionally necessary for time. Perhaps we judge that only worlds containing an A-series contain time, and, moreover, we judge this to be so even if the actual world does not contain an A-series. This latter view is what McTaggart (1908) and Gödel (1949) take our concept to be like. For McTaggart, it doesn't matter what the actual world is like: time exists if an A-series exists.[13]

Most of the studies we describe in the following chapter aim to determine, for a range of conditions, whether, for some, or all, of the folk concepts of time in the population we test, those conditions are conditionally necessary, unconditionally necessary, or not necessary at all. Our aim is to determine whether there are conditions under which people will, in general, deny that time exists. Suppose we find such conditions. Then it follows that the folk concepts of time are not immune to error.

---

[11] For instance, it is conditionally necessary that water is H_2O: given that the watery stuff is in fact H_2O, water is H20 in every world; but if the watery stuff is not H_2O, then it is not necessary that water is H_2O (perhaps it is not even possible).

[12] So described, A-theorists include presentists, growing block theorists and moving spotlight theorists. Defenders of some version of the A-theory, in various guises, include Craig (2000), Zimmerman (2008), Bourne (2006), Cameron (2015), Markosian (2004), Monton (2006), Prior (1968, 1967), Schlesinger (1994), Tallant (2012), and Tooley (1997). On such a theory there is a privileged moment that is present, and which moment is present, changes. Often, we will call such theories of time *dynamical* for just this reason.

[13] Since McTaggart concludes that the A-series is internally inconsistent, he ultimately concludes that time is impossible (and hence not actual).

4

The Sydney Time Studies

4.1. Introduction

In the previous chapters, we provided a broad summary of the results of the
Sydney Time Studies: a series of empirical studies carried out in order to
investigate certain aspects of the folk concept(s) of time, and we defended our
methodology. As promised, in this chapter we present the details of those
studies. This chapter, like the last, is entirely optional, and may be bypassed. It
is also designed to be largely standalone; those interested primarily in the
empirical details can focus only on this chapter. The chapter is structured as
follows. In §4.2, we say a bit more about our empirical methodology and, in
particular, how we aim to test for conceptual variability. In §4.3, we outline the
studies, explaining whether they confirm or disconfirm the hypotheses outlined in Chapter Three.

4.2. Conceptual Variability

How might we go about determining whether there is a univocal folk concept
of time in the population tested, or a multitude of folk concepts? The obvious
answer is that two people share a concept just in case their pattern of usage of
the concept reveals that their concept has the same necessary and sufficient
conditions for satisfaction. So, if we locate just one pattern of usage, then we
have just one folk concept. More than one pattern of usage would suggest
otherwise. But what is it for two concepts to have the same necessary and
sufficient conditions for satisfaction?

Let's return to our discussion in Chapter Three. We noted there that we
favour an approach on which an entire two-dimensional matrix reveals the
content of a concept. On such a view, x and y are the same concept, just in
case 'x' and 'y' share the same primary and secondary intensions: that is, just
in case the two-dimensional matrix for each is the same. On this view, we can
articulate a range of cases in which parties appear to share a concept, but do
not, and cases in which parties appear to employ two distinct concepts, but in

Out of Time: A Philosophical Study of Timelessness. Sam Baron, Kristie Miller, and Jonathan Tallant, Oxford University Press.
© Sam Baron, Kristie Miller, and Jonathan Tallant 2022. DOI: 10.1093/oso/9780192864888.003.0004

fact employ the same concept. We take this to be an attractive feature of the two-dimensional framework.

We can see this by considering a pair of examples. Consider, first, Hermione and Harry. When we present them with scenarios and ask them to consider these as counterfactual, both Harry and Hermione judge that there is time in all and only scenarios in which there is an A-series. So, both *appear* to judge that the presence of an A-series is both necessary and sufficient for time. It would be natural to conclude that they both share an A-theoretic folk concept.

Suppose, however, we ask each to imagine that the actual world has been discovered to be somewhat different than they supposed: it has been discovered to contain a B-series but no A-series. As it turns out, Harry judges that, in this eventuality, actually there *is* time. He judges that the presence of the B-series is sufficient for time, if that's all there actually is. Indeed, once Harry holds fixed that the actual world contains a B-series but no A-series, he then judges that any counterfactual world containing a B-series (but no A-series) contains time. By contrast, Hermione judges that if, actually, there is a B-series and no A-series, then actually there is no time—likewise for any counterfactual world.

We say that Harry and Hermione do *not* share a folk concept of time, despite initial appearances to the contrary. We say that Harry has a concept of time on which the presence of an A-series is *conditionally* necessary for time. Hermione has a concept on which the presence of an A-series is *unconditionally* necessary for time. They appear to share a concept of time because they *in fact* both make the same assumption about what the actual world is like. But the only way to see this is by presenting them with a scenario in which there is no A-series, and asking them to consider that scenario to be actual. This is what our methodology allows us to do.

Consider, next, Ron and Neville. Ron and Neville appear to employ different concepts of time. Ron judges that there is time in all and only counterfactual worlds in which there is an A-series, while Neville judges that there is time in all and only counterfactual worlds containing either an A-series or a B-series. So, Ron appears to judge that the presence of an A-series is necessary for time, while Neville appears to disagree.

But suppose that, as a matter of fact, Ron believes that our world is A-theoretic, while Neville believes that it is B-theoretic. Further, suppose that if we ask both Ron and Neville to imagine that it is has been discovered that there is only a B-series and no A-series, *both* are inclined to judge that there is actually time, and further, both are inclined to judge that there is time in counterfactual worlds that contain only a B-series. Then we will say that Ron and Neville share a folk concept of time. They merely appear to employ different concepts of time because they *disagree* about whether the actual world contains an A-series or not. Ron thinks it does; Neville thinks it does not. Since they both have concepts on which the presence of an A-series is conditionally necessary for there to be time, any disagreement they have about whether or not there is in fact such an actual A-series will translate into apparent disagreements about whether an A-series is necessary for time or not. But once we ask them to make the same assumption about the actual world, we find that their pattern of usage of the concept is the same.

One nice feature of our methodology is that it allows us to distinguish apparent conceptual differences that are merely differences in beliefs about how things are in the actual world, from genuine conceptual differences that manifest as differences in usage of the concept across scenarios considered as actual and counterfactual.

In all, then, in order to determine whether there are multiple folk concepts of time in the population tested, we need to determine whether participants' make different judgements about whether their concept is satisfied in scenarios *considered as actual,* and whether they make different judgements about whether their concept is satisfied in scenarios considered as counterfactual, *holding fixed which scenario is taken to be actual.* What we don't want to do is just look at people's judgements about what satisfies their concept in counterfactual scenarios, and conclude that if these judgements are different then they must be employing different concepts. It could be that this pattern of use in counterfactual scenarios is different only because participants are making different assumptions about what the actual world is like. As such, to determine whether participants share a folk concept of time, this is the methodology that we shall use.

4.3. The Studies

The Sydney Time Studies aimed to determine whether there is a univocal folk concept of time, and if there is, to determine at least some of the conditions that are necessary for that concept to be satisfied. If there is no such shared concept, then they aimed to isolate some shared necessary conditions.

The method, in line with the broad approach described in Chapters Two and Three, involves presenting participants with vignettes describing actual and counterfactual scenarios. Since talk of possible worlds is a philosophical conceit, we instead used the more familiar term 'universe'. Participants reading a vignette that described a counterfactual scenario were told it described a *parallel universe;* participants reading a vignette that described an actual scenario were told it describes *our universe.* So, in what follows, we will sometimes talk of participants' judgements about time in a particular universe; readers can feel free to switch out talk for universes for talk of scenarios or worlds.

Let's begin with our first hypotheses, which centre around whether (and in what respect) the presence of an A-series is necessary for our folk concept(s) to be satisfied. Recall the following trio of hypotheses:

Actual Time Dynamical Hypothesis: according to the folk, actual time is dynamical (i.e. it is characterised by an A-series).

The Conditional Dynamical Hypothesis: according to the folk concept of time, the presence of a dynamical A-series is conditionally necessary for time.

The Unconditional Dynamical Hypothesis: according to the folk concept of time, the presence of a dynamical A-series is unconditionally necessary for time.

In the first study (Latham, Miller and Norton 2019) we presented 600 participants with six vignettes. Three of the vignettes described dynamical models—presentism, the growing block and the moving spotlight—and three described non-dynamical models—eternalist B-theory, eternalist C-theory, and a D-theory model: a model according to which time is constituted by a D-series. The D-theoretic vignette describes a world in which there is no correct way to order events. Instead, according to the vignette, the universe described is like a deck of cards. Each card represents all the events that bear purely spatial relations to one another. In this universe, the only distance relations that exist are spatial distance relations. While there is a fact of the matter regarding the spatial relations between objects and events located on the same card, there is no fact of the matter as to the order of the cards. Any way of ordering the cards is just as good as any other way. Because of this, there is no fact of the matter about the distance relations between events on different cards. These vignettes were *not* time-neutral.[1] Having seen these vignettes participants were asked 'which of these Universes is most like our Universe?'.

---

[1] Recall that a vignette is time neutral when no temporal locutions are used to state the scenario at issue.

We found that a majority (66.3 per cent) of participants judge that one of the dynamical models of time is most like the actual world. We will say that these people have a *dynamical (or A-theoretic) theory of time,* and we will call such people dynamists (or A-theorists). A substantial minority (33.7 per cent) think a non-dynamical theory is most like the actual world. We will say that these people have a *non-dynamical theory of time,* and will call such people non-dynamists. Interestingly, though, participants were split between which dynamical and non-dynamical model they thought was most like our world. In this experiment, we found that 14.5 per cent of people chose the moving spotlight model, 17.4 per cent chose presentism, 34.3 per cent chose the growing block theory, 17.2 per cent chose the block universe model, 9.3 per cent chose the C-theory, and 7.3 per cent chose the D-theory model. Table 4.1 of these results is below.

These results go some way towards vindicating the Actual Time Dynamical Hypothesis: a majority of participants do judge that our world is A-theoretic. Do they, however, judge that the presence of an A-series is necessary (conditionally or unconditionally) for time?

Our second study, with 411 participants, aims to answer that question. In Latham, Miller and Norton (2020b), participants see a pair of vignettes, one of which describes a dynamical scenario, and one a non-dynamical scenario. We used time-neutral versions of two of the vignettes that were used as part of our first study. The non-dynamical scenario described a B-theoretic world (time-neutrally) and the dynamical scenario described a growing block world (time-neutrally). We chose a growing block world to represent a dynamical

Table 4.1. Frequency with which participants judged various philosophical models to be most like the actual world. Table 4.1 shows only of those participants that answered 2 out of 3 comprehension questions correctly.[2]

Dynamical			Non-Dynamical		
228 (66.3%)			116 (33.7%)		
Moving Spotlight	Presentism	Growing Block	Block	C-Theory	D-Theory
50 (14.5%)	60 (17.4%)	118 (34.3%)	59 (17.2%)	32 (9.3%)	25 (7.3%)

[2] It makes no difference to the reported result if we include those who do not comprehend the vignette they chose as being most like the actual world. When all responses are tallied, 332 (62 per cent) choose a dynamical theory of time and 203 (38 per cent) choose a non-dynamical theory of time.

theory of time since according to our first study it was overwhelmingly the most popular view about what time is actually like.

We will outline our methodology for this study in some detail in what follows, because a number of the studies that follow use the very same methodology, simply using different vignettes.

To gain information about what participants think our world is in fact like, in each experiment participants began by reading both vignettes. They were then asked which universe described in the vignettes is most like our universe. In this study, unlike the first, they are only given the two options to choose from. This sets the first index: whether people in fact think the actual world is dynamical or non-dynamical.

Participants then saw each vignette again (in random order). After seeing the first vignette, they were told that scientists have discovered that our universe is just like the universe described in the vignette. So, if participants first saw a vignette describing a dynamical universe, they were then told that this universe is just like our own, and *mutatis mutandis* if they first see a vignette describing a non-dynamical universe. Participants were then asked to agree, on a Likert scale of 1–7, with the claim (1) there is time in our universe. This corresponds to them being asked to consider some scenario as actual (namely the scenario described in the first vignette). We treated people who chose 5, 6 or 7 on the Likert scale to be responding that there is time, and those who chose 1 2, or 3 to be responding that there is no time. Those who chose 4, we took to be unsure. The process is then repeated for the second vignette. Jointly, this gives us information about what people say about whether there is actually time, conditional on them making certain discoveries about the actual world—namely that it is dynamical in a certain way, or that it is non-dynamical in a certain way.

All participants then saw both vignettes side by side. In one condition participants are told that the dynamical universe is just like the actual universe, and the non-dynamical universe is the parallel universe. In the other condition, they are told that the non-dynamical world is just like the actual universe, and the dynamical universe (presentist in one experiment, growing block in the other) is a parallel universe. In both conditions participants are then asked to respond, on a Likert scale, to the claim, (2), 'there is time in the parallel universe'. This corresponds to them being asked to evaluate a counterfactual world (the parallel universe) conditional on some particular world (dynamical or non-dynamical) having been discovered to be actual. Those in the first condition are asked to evaluate the parallel universe conditional on the actual world being dynamical, while those in the second condition are asked to evaluate the parallel universe conditional on the actual world being non-dynamical. We reproduce some of these results below (Table 4.2). For full details of this study see Latham, Miller and Norton (2020b). The somewhat smaller numbers of participants reported is the result of eliminating a large number of participants who did not fully understand the vignettes.

Since most of the tables that follow look like this one, we will work through some of this table so it is clear how to read the tables that follow.

The first group are those who judge that the dynamical world is most like our world (i.e. dynamists). The first row tells us what percentage of dynamists judge that there is time in a dynamical world considered as actual. Since this is what dynamists in fact think our world is like, we'd expect the percentage of dynamists who make this judgement to be quite high, and it is. The second row tells us what dynamists judge about whether there is time in a counter-factual growing block world, conditional on the actual world in fact being B-theoretic (*pace* what those dynamists in fact think it is like). The third row tells us what those dynamists judge about whether there is time in the actual world on the assumption that the actual world is B-theoretic And, finally, the last row tells us what dynamists judge about time in a counterfactual B-theoretic world, after having been told that, actually, our world is dynamical. The rest of the table is to be read the same way, except that in the other group we are asking about the judgements of non-dynamists: those who judge that our world is most like the scenario described in the B-theoretic vignette.

These results suggest that *most* participants have a concept which is such that the presence of an A-series is neither conditionally nor unconditionally

Table 4.2. Levels of agreement that there is time for different contexts given participants' belief about the actual world.

	%Yes	%No	%4	Mean	SD
Group: Growing Block is most like the actual world (*N* = 62)					
Actual Growing Block	88.7	6.5	4.8	5.89	1.38
Counterfactual Growing Block (Actual B-Theory)	85.5	11.3	3.2	5.71	1.43
Actual B-Theory	69.3	24.2	6.5	5.08	1.73
Counterfactual B-Theory (Actual Growing Block)	70.9	22.6	6.5	4.98	1.69
Group: B-Theory is most like the actual world (*N* = 79)					
Actual Growing Block	74.7	17.7	7.6	5.41	1.65
Counterfactual Growing Block (Actual B-Theory)	73.4	19	7.6	5.47	1.69
Actual B-Theory	94.9	3.8	1.3	6.18	1.05
Counterfactual B-Theory (Actual Growing Block)	83.6	13.9	2.5	5.71	1.50

necessary for time. So, neither the conditional nor unconditional dynamical hypotheses are vindicated. Notably, the discovery that there isn't an A-series in a world does not lead people to judge that there is no time in that world; this holds even if they are told that there is actually an A-series. Further, most people appear to judge that the presence of a B-series is sufficient for time, and this is so regardless of whether participants are supposing the actual world to contain an A-series or not, and regardless of whether they in fact think our world is dynamical.

Interestingly, though, these results suggest that there are two smaller sub-populations that do not share this view. We found that some dynamists judge that if it is discovered that the actual world does not contain an A-series, then this is the discovery that actually there is no time. Hence, these participants seem to have a concept on which an A-series is unconditionally necessary for time: for these participants, regardless of what the actual world is discovered to be like, if a world is not dynamical, it is not one at which time is real. This is the kind of population with which McTaggart would have found some measure of agreement.

Equally, it was found that some non-dynamists hold that if the actual world is *dynamical*, then *actually there is no time* (~20 per cent). This suggests that these non-dynamists deploy a concept on which not only is the presence of a B-series unconditionally necessary for time, but the presence of an A-series is *inconsistent* with the presence of time. For these participants, what is necessary and sufficient for time is that a world is B-theoretic, where, remember, a world is B-theoretic just in case it contains a B-series *and no A-series*. McTaggart would not find much to like in such company.

Taken together these results go some way towards suggesting that there is not a univocal folk concept of time amongst the population tested. Participants disagree about whether their concept of time is satisfied across scenarios considered as actual, and they disagree about whether their concept is satisfied across scenarios considered as counterfactual, even holding fixed how things are actually. This study, like the first study, tends to vindicate the actual time dynamical hypothesis, but neither the conditional nor unconditional dynamical hypotheses.

That brings us to our second cluster of hypotheses. Remember, they are as follows:

Actual Time Presentist Hypothesis: according to the folk concept of time, actual time is presentist (i.e. only present things exist, and which things are present, changes).

The Conditional Presentist Hypothesis: according to the folk concept of time, presentism is conditionally necessary for time.

The Unconditional Dynamical Hypothesis: according to the folk concept of time, presentism is unconditionally necessary for time.

We saw in our first study that while a majority of participants are dynamists, a minority of those think that our world is a presentist world. Indeed, of the total sample of participants, roughly the same percentage think our world is a presentist world, as think it is a B-theoretic block world. So, this suggests that the actual time presentist hypothesis is false. In order to test the second two hypotheses here, though, we need to look to a third study, also to be found in Latham, Miller and Norton (2020b).

Our third study was just like the second study, except that whereas in the second study people saw a growing block vignette as representing a dynamical scenario, in this study they saw a presentist vignette (again, made time-neutral). In every other respect the methodology was the same. 421 participants took part in the study.

Below Table 4.3 summarises some of the results. The remainder of the analyses can be found in Latham, Miller and Norton (2020b). Again, the data only include those who comprehended the vignettes.

The results here are very similar to those from our second study. Again, we fail to find that the presence of dynamism, this time in the form of presentism, is either conditionally or unconditionally necessary for time. Even amongst participants who think that our world is presentist, a majority still judge there

Table 4.3. Levels of agreement that there is time for different contexts given participants' belief about the actual world.

	%Yes	%No	%4	Mean	SD
Group: Presentism is most like the actual world ($N = 65$)					
Actual Presentism	83.1	13.8	3.1	5.60	1.56
Counterfactual Presentism (Actual B-Theory)	75.4	15.4	9.2	5.31	1.51
Actual B-Theory	60	32.3	7.7	4.69	1.86
Counterfactual B-Theory (Actual Presentism)	61.5	27.7	10.8	4.83	1.83
Group: B-Theory is most like the actual world ($N = 80$)					
Actual Presentism	70	25	5	4.89	1.68
Counterfactual Presentism (Actual B-Theory)	68.7	27.5	3.8	4.85	1.69
Actual B-Theory	88.7	8.8	2.5	5.68	1.27
Counterfactual B-Theory (Actual Presentism)	87.4	8.8	3.8	5.51	1.31

to be time in non-presentist scenarios that contain only a B-series, regardless of whether those scenarios are being considered as actual or counterfactual. As such this study also suggests that there is more than one folk concept present in the population.

While most participants do not think that presentism is necessary (conditionally or unconditionally) for their concept to be satisfied, as in Study Two we see that there are participants who do not respond in this manner: ≈40 per cent of presentists judge that presentism is unconditionally necessary for time, and ≈30 per cent of B-theorists judge that presentism is unconditionally *inconsistent* with time. This pattern of responses suggests that there are multiple folk concepts in the population tested.

This brings us to the third cluster of hypotheses, which centre around the role of a B-series. Recall they are as follows:

Actual Time B-Theory Hypothesis: according to the folk concept of time, actual time is B-theoretic (i.e. it is characterized by a B-series but not an A-series).

The Conditional B-series Hypothesis: according to the folk concept of time, the presence of a B-series is conditionally necessary for time.

The Unconditional B-series Hypothesis: according to the folk concept of time, the presence of a B-series is unconditionally necessary for time.

Again, our very first study speaks to the actual time B-theory hypothesis. That hypothesis is false. While a small percentage of participants tested do think our world is B-theoretic, most think it is some other way.

To determine whether the conditional or unconditional B-series hypotheses are true, however, we needed to run a fourth study. In this study, we used the same methodology as that just described. Indeed, we used the same time-neutral B-theory vignette in this study. But we replaced the dynamical vignette with a time-neutral vignette that described a world containing only a C-series and no B-series or A-series (that is, we described a C-theoretic world). We had 396 participants, but a large number of these were excluded for failing to comprehend the vignettes. However, analyses of the complete data set show no statistically significant difference between this smaller data set, and the complete set. For the complete analyses see Latham, Miller and Norton (2020a). The results of this study are below, in Table 4.4.

This study failed to vindicate either the conditional or unconditional B-series hypothesis. A majority of participants judge there to be time in

Table 4.4. Levels of agreement that there is time for different contexts given participants' belief about the actual world.

	%Yes	%No	%4	Mean	SD
Group: B-Theory is most like the actual world ($N = 61$)					
Actual B-Theory	91.8	1.6	6.6	5.70	0.92
Counterfactual B-Theory (Actual C-Theory)	88.5	6.6	4.9	5.51	1.11
Actual C-Theory	73.8	16.4	9.8	5.13	1.52
Counterfactual C-Theory (Actual B-Theory)	67.2	24.6	8.2	4.90	1.54
Group: C-Theory is most like the actual world ($N = 34$)					
Actual B-Theory	85.3	11.8	2.9	5.35	1.41
Counterfactual B-Theory (Actual C-Theory)	94.1	5.9	0	5.62	1.05
Actual C-Theory	85.3	11.8	2.9	5.41	1.42
Counterfactual C-Theory (Actual B-Theory)	85.3	8.8	5.9	5.35	1.18

scenarios (considered as actual or counterfactual) in which there is a C-series but no B-series.

We do, though, see a difference between B-theorists (those who judge that the B-theoretic world is most like our world) and C-theorists (those who judge that the C-theoretic world is most like our world) in the following way: *more* B-theorists judge that there is *no* time in the counterfactual C-theoretic world, than do C-theorists. Moreover, *more* B-theorists judge that there is time in an actual C-theoretic world compared to a counterfactual C-theoretic world, conditional on the actual world being B-theoretic. So, it does look as though some B-theorists have a concept in which the presence of a B-series is conditionally necessary for time, while most B-theorists (and C-theorists) have a concept in which the presence of a B-series is neither *conditionally* nor unconditionally necessary for time. Again, this supports our earlier findings that there are multiple folk concepts in the population tested.

That brings us to our fourth cluster of hypotheses:

Actual Time C-Theory Hypothesis: according to the folk concept of time, actual time is C-theoretic (i.e. it is characterised by a C-series but not an A-series or a B-series).

The Conditional C-series Hypothesis: according to the folk concept of time, the presence of a C-series is conditionally necessary for time.

The Unconditional C-series Hypothesis: according to the folk concept of time, the presence of a C-series is unconditionally necessary for time.

Returning again to our first study, we see that, unsurprisingly, the actual time C-theory hypothesis is false. A very small percentage of the total population tested think our world is C-theoretic.

In order to test the latter two hypotheses, we ran a further two studies: Studies Five and Six. These studies are reported in Latham and Miller (2020a). They had a very similar methodology to the studies we have already described. One difference, though, was that we presented participants with *three* (time-neutral) vignettes at the beginning of each study: a dynamical vignette (growing block), a C-theoretic vignette a D-theoretic vignette. In both studies five and six, participants see all three vignettes, and are asked which they think is most like our world. This allows us to more finely divide up participants into A-theorists, C-theorists and D-theorists (those who think that our world is D-theoretic, as described above).

As in the previous studies we described, each participant saw two vignettes. The methodology then follows the methodology we have already described. In the first study participants see a D-theoretic vignette and a growing block vignette, and in the second study they see a D-theoretic vignette and a C-theoretic vignette. So, in both studies participants are responding to one vignette that describes a scenario in which there is at least a C-series, and a vignette that describes a scenario in which there is no C-series. However, in the first study one vignette has a C-series in virtue of having an A-series, and in the latter, it merely has a C-series.

Table 4.5, below, presents the descriptive data of Study Five. This table includes both a t-value and a p-value. The combined p-value and t-value tell us whether the mean differs significantly from 4. In particular a p-value of <0.05 shows that the t-value is significant: that is, the mean does differ significantly from 4.

While these results do not support the conditional C-series hypothesis, they *do* support the unconditional C-series hypothesis (though we return to this issue after we report the next study). Overall, a majority of participants judge that the presence of a C-series is unconditionally necessary for time.[3]

This study also tends to support the idea that there are multiple folk concepts present in the population tested. First, we see that a higher percentage of D-theorists (and C-theorists) judge there to be time in the D-theoretic

---

[3] At first glance that might not seem quite right: after all, nearly 50 per cent of D-theorists don't think that the presence of the C-series is necessary (conditionally or unconditionally) for time. But remember that D-theorists are a very small percentage of the total population. Most participants in this study were dynamists, and most dynamists thought that the presence of a C-series was unconditionally necessary for time.

Table 4.5. Levels of agreement that there is time in scenarios taken as actual or counterfactual. In parentheses we note how the actual world is taken to be when making the judgement about the counterfactual world.

	%Yes	%No	%4	Mean	SD	t-value	p-value
Group: Growing Block is most like the actual world ($N = 109$; 73.2%)							
Actual Growing Block	92.7	5.5	1.8	5.98	1.24	16.690	<.001
Counterfactual Growing Block (Actual D-Theory)	84.4	12.8	2.8	5.51	1.55	10.201	<.001
Actual D-Theory	30.3	65.1	4.6	3.26	1.76	−4.406	<.001
Counterfactual D-Theory (Actual Growing Block)	36.7	59.6	3.7	3.56	1.89	−2.429	.017
Group: C-Theory is most like the actual world ($N = 25$; 16.8%)							
Actual Growing Block	88	12	0	5.48	1.26	5.862	<.001
Counterfactual Growing Block (Actual D-Theory)	64	24	12	4.96	1.59	3.012	.006
Actual D-Theory	52	48	0	3.92	2.06	−0.194	.848
Counterfactual D-Theory (Actual Growing Block)	36	56	8	3.64	1.85	−.975	.339
Group: D-Theory is most like the actual world ($N = 15$; 10.1%)							
Actual Growing Block	66.6	26.7	6.7	4.93	1.79	2.018	.063
Counterfactual Growing Block (Actual D-Theory)	53.3	46.7	0	4.13	2.07	.250	.806
Actual D-Theory	46.7	33.3	20	4.33	1.92	.774	.511
Counterfactual C-Theory (Actual Growing Block)	46.7	33.3	20	4.47	1.69	1.073	.301

world, than do the dynamists (though we did not perform comparative statistical analyses to see if this difference was significant due to the small sample sizes of participants in those last two groups, once we removed participants who did not comprehend the vignettes). Second, we see that, amongst D-theorists, there is practically a 50/50 split between those who think there is time in a D-theoretic world, and those who think there is not, regardless of whether it is considered as actual. Third, although a clear majority of dynamists judge that there is no time in a D-theoretic world, close to 30 per cent think there is time in such a world (considered as actual or counterfactual). In all, this suggests that there are a number of sub-populations who have concepts of time that are quite different: at one end of the spectrum we have those who think that the presence of even a C-series is not necessary (even conditionally) for time, and at the other end are those who think that the presence of a C-series is unconditionally necessary for time.

Bearing that in mind, let's consider our sixth study. This study is like the one we just described, except that after seeing all three vignettes participants then see a different pair of vignettes: a D-theoretic vignette, and a C-theoretic vignette. The descriptive data from Study Six is in Table 4.6 below.

The results of this study are similar, in crucial respects, to those of Study Five. Again, we find that most dynamists hold that a C-series is unconditionally necessary for time. Contrary to Study Five, however, we find that a *majority* of both C-theorists and D-theorists judge that there is time in a D-theoretic world considered as actual or counterfactual. Hence, a majority of these participants appear to have a concept on which the presence of a C-series is neither conditionally nor unconditionally necessary for time. On reflection, however, given the very small number of C- and D-theorists in each study, it only takes a few participants to respond differently in one study compared to another, and that small number will be enough, in a small sample size, to

Table 4.6. Levels of agreement that there is time in scenarios taken as actual or counterfactual. In parentheses we note how the actual universe is taken to be when making the judgement about the counterfactual scenario.

	%Yes	%No	%4	Mean	SD	*t*-value	*p*-value
Group: Growing Block is most like the actual world ($N = 91$; 62.3%)							
Actual C-Theory	50.5	42.9	6.6	4.07	1.73	0.363	.717
Counterfactual C-Theory (Actual D-Theory)	48.3	46.2	5.5	3.98	1.85	−0.113	.910
Actual D-Theory	28.6	68.1	3.3	3.15	1.75	−4.611	<.001
Counterfactual D-Theory (Actual C-Theory)	30.8	63.7	5.5	3.25	1.76	−4.047	<.001
Group: C-Theory is most like the actual world ($N = 39$; 26.7%)							
Actual C-Theory	87.1	10.3	2.6	5.69	1.24	8.532	<.001
Counterfactual C-Theory (Actual D-Theory)	82	15.4	2.6	5.28	1.56	5.148	<.001
Actual D-Theory	69.2	30.8	0	4.72	2.06	2.172	.036
Counterfactual D-Theory (Actual C-Theory)	76.9	20.5	2.6	4.87	1.81	3.010	.005
Group: D-Theory is most like the actual world ($N = 16$; 11.0%)							
Actual C-Theory	68.7	12.5	18.8	4.88	1.67	2.098	.053
Counterfactual Growing Block (Actual D-Theory)	68.7	31.3	0	4.44	1.83	0.959	.353
Actual D-Theory	81.2	18.8	0	5.50	1.75	3.426	.004
Counterfactual D-Theory (Actual C-Theory)	87.5	12.5	0	5.56	1.26	4.948	<.001

change the mean participant response. This means that we cannot speak to whether a majority of C-theorists or D-theorists have a concept on which the presence of a C-series is unconditionally necessary, or a majority have a concept on which the presence of a C-series is not necessary at all. What seems clear is that *some* have one concept, and some have the other, but because there are so few of these people in the population, these two experiments cannot tell us any more than that.

In all, though, the totality of results across Studies Five and Six also suggests that there are multiple folk concepts in the population. Further, since dynamists were by far the largest group of participants, both studies support the unconditional C-series hypothesis: most people do judge that the presence of a C-series is necessary for time, and they judge this regardless of whether the actual world contains a C-series or not.[4]

That brings us to our fifth hypothesis: the dual functionalist hypothesis.

The Dual Functionalist Hypothesis: according to the folk concept of time, time is whatever it is that plays *both* the role of grounding our temporal seemings and the role of grounding causation and change.

To test this hypothesis, Latham and Miller (2021) ran a seventh study with 510 participants. In this study participants read five vignettes, each of which describes a counterfactual universe. As before, at the end of each vignette participants were either told that our universe is *just like* the universe described in the vignette, or that our universe *differs from* that universe in certain ways, which we will specify shortly.

Some participants were presented with a vignette in which there is *counterfactual tracking*. Individuals in that world are described as *tracking*, with their temporal phenomenology, that thing that grounds causation and change in that world. Participants who saw this vignette were told that the actual world is *like* the counterfactual world described, in that the same thing that grounds causation and change in the counterfactual world obtains in the actual world. However, those participants were then *either* told that in the actual world people's phenomenology tracks what actually grounds causation and change (*actual tracking*) or they are told that actual people's phenomenology fails to track what actually grounds causation and change (*actual*

---

[4] As we noted in Chapter Two, however, it remains to be seen whether in fact it is the unconditional C-series hypothesis that is confirmed, or something very like that hypothesis. We noted in Chapter Two that the vignettes we show participants do not distinguish between scenarios that lack or contain a C-series, and those that lack or contain some other feature.

non-tracking). In particular, in the actual non-tracking condition participants are told that our actual temporal phenomenology:

> is not caused in the same way as the temporal phenomenology in the counterfactual tracking world. Instead, we have the temporal phenomenology we do because an evil demon directly creates these experiences in us. The reason it seems to us that events happen in a particular order is because the demon creates experiences in which this seems to be the case. The reason it seems to us as though events have a certain duration, is because the demon creates experiences in which this seems to be the case. The evil demon is responsible for all of us having the temporal phenomenology we do.

To be clear, then, in the actual non-tracking condition our actual temporal phenomenology is discovered not only to *not* be tracking the thing (a C-series, or an A-series) that grounds change and causation, but, in addition that phenomenology is caused by an evil demon.

Hence, these are either scenarios in which there is counterfactual tracking and actual tracking, or counterfactual tracking and actual non-tracking.

In addition, what is being tracked in the counterfactual world can either be similar to what people in fact think they are tracking (similar) or dissimilar from what they in fact think they are tracking (dissimilar). Here, based on the very first study we described in this chapter, regarding what people think time in the actual world is like, we supposed that if the growth of a growing block grounds causation and change actually, then this is similar to what (most) people in fact think grounds causation and change, and is similar to what they suppose their temporal phenomenology to be tracking. By contrast, if what grounds causation and change actually, and what is being tracked by our actual temporal phenomenology, is just the presence of a C-series ordering of events, then this is dissimilar to what people in fact suppose their temporal phenomenology to be tracking.

Bearing this in mind, the counterfactual world with tracking can be a world in which the temporal phenomenology of individuals in that world is tracking something similar to what actual people think their own phenomenology is tracking, (the accretion of a growing block world) and it can be that actual people's phenomenology is tracking the very same thing (counterfactual tracking, actual tracking, similar), or it can be that people's actual phenomenology is failing to track that thing and is instead being produced by a demon (counterfactual tracking, actual non-tracking, similar). Or it can be that in counterfactual worlds, what an individual's phenomenology is tracking is dissimilar to what actual people think their own phenomenology is tracking, (a C-series order of events), and it can either be that in the actual world our phenomenology is tracking that thing (counterfactual tracking, actual tracking, dissimilar) or that in the actual world, our phenomenology is failing to track that thing (counterfactual tracking, actual non-tracking dissimilar) and is instead being produced by a demon.

Finally, some participants were presented with a vignette in which there is no counterfactual tracking—i.e. people's phenomenology in the counterfactual world fails to track what grounds causation and change because there is nothing in that world that grounds causation and change. In this condition participants are told that the actual world is just like this, and so in the actual world people's phenomenology fails to track what grounds causation and change, since there is nothing in the actual world that grounds causation and change (counterfactual non-tracking, actual non-tracking).

In all, then, participants were presented with one of the following five vignettes.

1. There is something that grounds causation and change and that something is similar to what individuals expect to ground their temporal phenomenology, and their actual phenomenology is grounded by that thing (counterfactual tracking; actual tracking; similar)

2. There is something that grounds causation and change and that thing is similar to what individuals expect to ground their temporal phenomenology, but their actual phenomenology fails to track that thing (counterfactual tracking; actual non-tracking, similar)

3. There is something that grounds causation and change and that something is dissimilar to what individuals expect to ground their temporal phenomenology, and their actual phenomenology tracks that thing (counterfactual tracking; actual tracking; dissimilar)

4. There is something that grounds causation and change and that something is dissimilar to what individuals expect to ground their temporal phenomenology, and their actual phenomenology fails to track that thing (counterfactual tracking; actual non-tracking, dissimilar)

5. Nothing grounds causation and change (because there is no causation and change) and hence there is nothing for actual temporal phenomenology to track (counterfactual non-tracking, actual non-tracking).

Below is a table of some of the results (Table 4.7).

Table 4.7. Levels of agreement that there is time in the actual world.

	%Yes	%No	%4	Mean	SD
Actual Tracking, Counterfactual Tracking, Similar (N = 105)	91.4	2.9	5.7	6.01	1.11
Actual Non-Tracking, Counterfactual Tracking, Similar (N = 105)	81.9	10.5	7.6	5.82	1.52
Actual Tracking, Counterfactual Tracking, Dissimilar (N = 102)	95.1	2.0	2.9	6.28	0.99
Actual Non-Tracking, Counterfactual Tracking, Dissimilar (N = 96)	87.4	6.3	6.3	5.82	1.31
Actual Non-Tracking & Counterfactual Non-Tracking (N = 97)	86.6	8.2	5.2	5.80	1.37

If the dual functionalist hypothesis is correct, then we should find that a majority of people judge that there is not actually time, if what our actual temporal phenomenology tracks does not also ground causation and change.

A majority of people in fact judge that there *is* time even when what grounds actual temporal phenomenology does not ground causation or change: that is, even when our temporal phenomenology is the result of our experiences being caused by a demon. 81.9 per cent of participants judge that there is actually time when actual temporal phenomenology is failing to track the growth of a growing block, and 87.4 per cent of participants judge that there is actually time when actual temporal phenomenology is failing to track a C-series. Overall, in every condition, a substantial majority of people think there is time in the actual world regardless of whether our phenomenology is being caused by an evil demon, as opposed to tracking a C-series or an A-series.

Secondly, if dual functionalism were correct, then a majority of people would judge that there is *no* time in the actual world, if there is no causation and change, and hence nothing grounds those things. This is not what we found. Not only did 77.3 per cent of participants judge that there is time in a counterfactual world where there is no causation and change, but 86.6 per cent judged that there is actually time, even if there *is* no causation and change and hence our temporal phenomenology is not tracking the thing that grounds causation and change.

In all, these results suggest that the folk concept of time is not a dual functionalist one. It made basically no difference whether what our temporal phenomenology tracks also grounds causation and change, to people's judgements about whether there is time in a scenario.

These results make it very unlikely that *any* version of functionalism that is like dual role functionalism can be right, even a version that spells out the second role quite differently. That is because participants were still inclined to say that there was time in the condition in which their temporal phenomenology is brought about by an evil demon, and where there is no causation and change (and hence the thing their temporal phenomenology tracks does not ground causation or change).

This finding is, however, consistent with the much weaker version of functionalism that we called Seeming Role Functionalism.

Seeming Role Functionalism: time is whatever it is that plays the role of grounding our temporal seemings.

This is the view on which time is just whatever it is that grounds our having the temporal phenomenology we do.

Our results provide some support for this version of functionalism. Having said that, these results are also consistent with the folk having a concept that is not functionalist at all, so that time is in no way what grounds our temporal phenomenology. These results are consistent with people having a fairly minimal non-functional concept.

The hypothesis that people have a very minimal non-functional concept is consistent with the results we find in Conditions 1 to 4. In all these conditions, there is something that could plausibly be thought to be temporal structure (a growing block A-series or a C-series) and our results suggest that people are indifferent to whether or not their phenomenology is tracking that thing. If they have a concept whose content does not mention temporal phenomenology, this is to be expected.

Nevertheless, this does not entirely explain the results we found in Condition 5, in which there is nothing in the actual (or counterfactual) world that includes even a one-dimensional sub-structure of ordered instants. Participants' general willingness to judge that there is time in the actual world, if it is like that, suggests that the concept in question must be quite minimal in its requirement.

In light, however, of our earlier results we are inclined to say that overall, the evidence suggests that people judge that the presence of a C-series (or some other feature missing from the D-theory vignettes) is unconditionally necessary for time. We have found very little evidence that people care what their temporal phenomenology is tracking, and considerable evidence that people judge that in worlds like the D-theory world which lacks a C-series, there is no time. That gives us reason to reject functionalist views, at least of the sort we have considered here (both dual role and seeming views) in favour of the view that what is necessary for the folk concept to be satisfied is some fairly minimal metaphysical structure.

We come now to the last hypothesis: Tallant's hypothesis.

The Unconditional Presence Hypothesis: according to the folk concept of time, present-tensed truths are unconditionally necessary for time.

Latham and Miller (2020b) tested this hypothesis as part of a study that also returned to an earlier hypothesis: namely that the presence of a C-series is unconditionally necessary for time. To do so, we introduced our 512 participants to what we call a one-slice vignette: a vignette that describes a scenario in which there is a single instant: a single 'slice' of reality, where events/objects located on that slice bear spatial, but not temporal, relations to one another, and where that slice does not undergo dynamical change (i.e. change in the total set of facts).

One way to conceptualise such a scenario is a 'stopped presentist' world: an unchanging three-dimensional slice of reality. Another reason to appeal to such scenarios is to test the unconditional presence hypothesis. After all, what *exactly* would it be for a world to lack present-tensed truths? To put it another way, what sort of scenario could we describe to participants, in which there are no present-tensed truths, in order to see whether they take such truths to be necessary for time? We wanted to pull apart the presence of present-tensed truths from the presence of the A-, B-, or C-series: after all, if there are such truths just when there is at least a C-series, then it might be true that such truths are necessary for time, just because a C-series is necessary for time.

To do so, we focused on vignettes describing one-slice scenarios. Very roughly, some vignettes describe one-slice scenarios in which participants are told that there are present-tensed truths, but no past or future-tensed truths; and some describe one-slice scenarios in which participants are told there are neither present- nor past- or future-tensed truths. With respect to these vignettes, two questions naturally arise: first, do people think there is time in any of the one-slice scenarios, and, second, if they do, are they more likely to think there is time in the one-slice scenarios in which there are present-tensed truths, as the unconditional presence hypothesis entails?

Participants in this study saw one of three vignettes. Each vignette describes a one-slice world that participants are told is a complete physical duplicate of our world at an instant in 1914.

In each vignette participants are presented with three particular tensed claims: a past-tensed claim, a present-tensed claim and a future-tensed claim. Each of these claims is true, *in our world*, in that instant in 1914.

Rather than stipulating the truth or falsity of these claims in the universe described, we tell participants one (and only one) of the following.

1. That scientists in our universe are *sure* that all three claims are false (absence of present-tensed truths condition).
2. That scientists are *sure* that the past- and future-tensed claims are false, and are also *sure* that the present-tensed claim is true (present-tensed truth condition).
3. That scientists are *sure* that the past- and future-tensed claims are false, but are *uncertain* whether the present-tensed claim is true (uncertainty condition).

Participants were then asked to agree/disagree with two statements. The first statement is that the present-tensed claim is true in the universe described in the vignette. Responses to this question allow us to determine whether the participants believe that the scientists are right about the presence, or absence, of present-tensed truths in the one-slice world. We can then correlate participants' beliefs about whether or not there are present-tensed truths in a one-slice world, with their judgments about whether or not the world contains time. Participants are then asked to evaluate whether or not there is time in the world described by the vignette as with the other studies conducted. The results are summarized below in Table 4.8. Again, participants who did not comprehend the vignettes were eliminated from the analyses.

The above table shows that people's mean judgment level of agreement regarding whether there is time in each scenario is around 4: they neither agree, nor disagree, that there is time. However, that was not because people were uncertain about whether there is time in universe A, B, or C. Rather, it's because they are divided between those that think there is time, and those that think there is no time or are uncertain whether there is time (see Figure 4.1 below).

In the present-tensed condition, participants were significantly more likely to agree that there were present-tensed truths than they were to agree that there were present-tensed truths in the absence of present-tensed truths condition. Further, participants in the uncertainty condition were significantly more likely to agree that there were present-tensed truths, than were participants in the absence of present-tensed truth condition. Overall, then, there was

Table 4.8. Levels of agreement to 'there is time in universe A/B/C' and to 'there are present-tensed truths in A/B/C'.

Question and Condition	%Yes	%No	%4	Mean	SD
"There is time in universe A/B/C."					
Uncertainty Condition (Universe A)(N = 73)	53.4	42.5	4.1	3.99	1.95
Present-Tensed Truth (Universe B) (N = 74)	60.8	33.8	5.4	4.28	2.08
Absence of Present-Tensed Truth (Universe C) (N = 72)	48.6	43.1	8.3	4.06	1.86
'Sentence (c), "presently, there are physical properties," is true in universe A/B/C.'					
Uncertainty Condition (Universe A)(N = 73)	82.2	12.3	5.5	5.41	1.32
Present-Tensed Truth (Universe B) (N = 74)	97.2	1.4	1.4	5.97	0.89
Absence of Present-Tense Truth (Universe C) (N = 72)	57.0	33.3	9.7	4.24	2.05

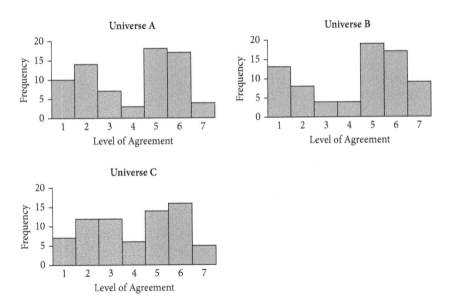

Figure 4.1. Histograms showing people's levels of agreement to the statement 'There is time in universe A/B/C.' in the three experimental conditions (universe A, B, and C).

a robust correlation between participants' being in a present-tensed truth condition, and judging that there are present-tensed truths, and between participants being in an absence condition, and judging that there are no present-tensed truths. Despite this, we found no association between people's present-tensed truth judgments and their judgments about whether there is time in the scenario in question. The presence of present-tensed truths (or participants' beliefs in their presence) made no difference to whether they judged there to be time in a scenario. Hence, we found no evidence in favour of the unconditional presence hypothesis.

Interestingly, in all three conditions roughly 50 per cent of participants (53.4 per cent uncertainty condition; 60.8 per cent present-truths condition; 48.6 per cent no present-tensed truths) judged that there was time. In all three conditions the mean level of agreement that there was time was around 4 (the mean varied from 3.99 to 4.28). But, as noted, this was not the result of a large number of participants being unsure whether or not there was time in the condition in question. Instead, we found a bi-modal distribution, with only 5.9 per cent of participants giving a response of 4 (neither agreeing, nor disagreeing, that there was time in the condition in question). As we can see from the histograms, the population is split into two across each condition: those who judge there is time in that condition, and those who judge that there is not.

4.4. Conclusion

The Sydney Time Studies provide evidence for three broad conclusions. First, they provide evidence that many, but not all, people are dynamists, and that very few of these dynamists have a dynamical (i.e. A-theoretic) concept of time: very few take dynamism to be conditionally or unconditionally necessary for time.

Second, they provide evidence for substantial conceptual variation in folk concepts of time. People's concepts vary from those that appear to be ones on which the presence of an A-series is unconditionally necessary for time, to those which take the B-series to be unconditionally necessary (but not the A-series), to those which take the C-series to be unconditionally necessary (but not the B-series), to those which take none of these to be necessary (this last class of people is very small indeed).

Third, they provide evidence that there are discoveries that people could make, which would lead them to conclude that time does not exist. Exactly what these discoveries might be, depends on the particular concepts that people are employing. By far the biggest effect we saw, though, was for the D-theory vignettes. When presented with this kind of case most dynamists deny that time exists, and dynamists constitute around two-thirds of the population. Many C-theorists are also willing to give up the existence of time in the face of this kind of discovery. Since the main feature of the

D-theory vignette is the lack of any temporal ordering or metric structure of the kind found in the C-series, it is reasonable to suppose that the absence of the C-series is what would lead people to deny that time exists.

This brings us to the end of Part One. In Part One, we have argued that the folk concept of time is not immune to error. There are discoveries that would seem to lead people to deny the existence of time. Of course, even if there are hypothetical discoveries that would lead people to deny that time exists, it may still turn out that such discoveries are incompatible with what we know about the world. We now need to argue that this is not so. That is our goal in Part Two. This sets the scene for Part Three of the book in which we consider the final reason to suppose that the loss of time is unthinkable: namely that it would undermine agency. We use results from Part Two of the book to argue that there can be agency without time.

Note that in Part Two, we use the discovery that there is no C-series as our stalking horse, since vignettes in which there is no C-series had, by far, the largest impact on people's willingness to deny that time exists. For reasons already discussed, however, we have to be a bit careful about taking the C-series to be unconditionally necessary for the existence of time, and so we will be sensitive to this complexity as we go.

PART II

5

A Quick Argument for Timelessness

5.1. Introduction

In Part One, we outlined a range of empirical results concerning the folk concept of time. We found that there is no single folk concept of time, but instead a range of folk concepts. We also found that there are scenarios that lead a majority of people to deny that time exists. Specifically, scenarios in which a C-series goes missing appear to yield a strong tendency to conclude that time does not exist. This is not yet enough to establish that time might not exist (as is our aim). For one might grant that folk concepts of time are not immune to error, and thus there are some hypothetical scenarios that people would judge are timeless, but claim that such discoveries are incompatible with what we know about the world from our best science.

Our goal in Part Two of the book is to argue against such a claim by making a case for the epistemic possibility of scenarios that lack a C-series. Roughly put, there are physical theories in which the C-series seems to be eliminated, and so the C-series might not exist, for all we know. We then present a challenge to this quick argument based on emergence. According to a number of philosophers and physicists, spacetime will be emergent within the theories at issue. The emergence of spacetime, it is also argued, implies the existence of spacetime, which in turn implies the existence of the C-series. Thus, these physical theories challenge the *fundamentality* but not the *existence* of a C-series. So, these views pose no special challenge to the satisfaction of our folk concepts of time.

In the rest of this chapter, we respond to a particular version of this challenge, based on a specific notion of emergence that we call *theoretical emergence*. Theoretical emergence, very roughly, is a deductive relationship between theories. We argue, however, that theoretical emergence need not have any ontological implications. The only way in which the challenge could succeed is if spacetime must emerge in a metaphysically weighty fashion. We go on to consider a more metaphysically weighty picture of emergence, and thus a more pressing version of the emergence challenge, in subsequent chapters.

Out of Time: A Philosophical Study of Timelessness. Sam Baron, Kristie Miller, and Jonathan Tallant, Oxford University Press. © Sam Baron, Kristie Miller, and Jonathan Tallant 2022. DOI: 10.1093/oso/9780192864888.003.0005

5.2. The Possibility of Timelessness

According to the standard understanding of one of our best scientific theories—the general theory of relativity (GR)—spacetime (or spatiotemporal relations) exist. The existence of spacetime thus *appears* to be at odds with the scenarios that lead people to reject the existence of time. After all, it is a feature of those scenarios that they lack temporal ordering and temporal metric structure (ordering and metric enough for, at least, a C-series). Spacetime seems unlikely to lack these features.

However, recent developments in physics have cast doubt on the existence of spacetime;[1] in particular, developments in the field of quantum gravity. Producing a quantum theory of gravity is the project of unifying our two best physical theories: GR and the standard model of particle physics, captured by the various quantum field theories.[2] Unification requires giving a quantum account of the gravitational field.[3] We don't have anything like an agreed-upon theory of QG yet. Part of the problem is that there is a dearth of empirical evidence that might confirm or disconfirm a given approach to QG. It has long been thought that theories of QG make predictions at a scale that is well below our current capacity to probe. While there has been some recent work on QG suggesting that the predictions of various theories at

---

[1] Timelessness has also been defended within philosophy. See McTaggart (1908) for the locus classicus, but see Earman (2002) for a modern take.

[2] Not everyone interprets the quantum gravity project in this way. For some, the goal of producing a quantum theory of gravity is nothing more than the project of providing a quantum account of the gravitational field. On its own, this may not result in the unification of general relativity with quantum theory, and perhaps we should manage our expectations in this direction accordingly. See Crowther (2016) for discussion.

[3] There are a number of motivations for developing such a quantum theory of gravity, and we cannot do justice to them all here. We can, however, outline three key motivations in this direction. First, gravity is thought to be one of four fundamental forces, the other three being: the weak nuclear force, the electromagnetic force and the strong nuclear force. To date, all three of the other forces have been captured in quantum terms. There is an expectation that gravity can be captured in the same terms, which in turn will lead toward a single unified quantum theory of all four of the fundamental forces. Second, and relatedly, quantum theory appears to be fully universal, insofar as it suggests that all dynamical aspects of the universe have quantum properties. Now, in GR, gravitational fields are the dynamic interaction between mass-energy and spacetime. This is captured by the coupling of the stress-energy tensor (which describes the mass-energy distribution across the universe) with the metric tensor (which describes the metric structure of the universe) in the field equations—the fundamental equations of the theory. Given that gravity is a dynamical entity, and as noted above quantum theory appears to be fully universal, there is an expectation that we should be able to give a quantum account of it. Third, as Butterfield and Isham (1999) note, general relativity gives rise to singularities in which spacetime curvature and matter density take infinite values. Arguably, the infinite values for curvature and density are mathematical pathologies of the field equations. There is hope that these pathologies will be removed by a quantum account of gravity. The basis for this hope appears to be the suggestion of a length-scale cut-off revealed in nature. Roughly, spacetime will, at some level, be *discrete* in a quantum account of gravity. Since it is the continuum which seems to enable the pathologies of general relativity, discreteness brings with it the prospect for panacea.

small scales might 'percolate up' to scales that we currently have the capacity to probe, work in this area is still in its infancy.[4]

Despite the absence of any agreed-upon theory, a number of approaches to QG have been proposed. The best-known approach to QG is string theory, according to which the fundamental components of reality are tiny one-dimensional strings that vibrate in up to eleven dimensions. The chief alternative to string theory (at least, in terms of number of physicists involved) is Loop Quantum Gravity (LQG), according to which reality is fundamentally a lattice-like structure, constituted by discrete 'chunks' that are 'woven' together. Another approach to QG is the Canonical Quantum Gravity program (CQG), which applies standard quantization techniques from quantum field theory to the gravitational field, in order to produce a quantum account of gravity (roughly in the mould of the standard model of particle physics). These are by no means the only available approaches to QG. Other approaches include causal set theory (Dowker 2013, Dowker 2014, Rideout and Sorkin 1999), the asymptotic safety approach (Benedetti, Machado, and Saueressig 2009), the causal triangulation approach (Ambjørn, Jurkiewicz, and Loll 2001, 2004) and the emergent gravity program (Hu 2009).

With respect to a number of these approaches to QG, both philosophers and physicists maintain that spacetime does not exist in some sense (we will explore some of these theories in a bit more detail in a moment). Of course, if spacetime does not exist according to these approaches and some other temporal structure does—like the A-, B-, or C-series—then developments in QG merely serve to reinforce the idea that the discoveries that would lead people to reject the existence of time are incompatible with what we know about the world from science (since those discoveries all require, at a minimum, the absence of the A-, B-, and C-series).

Recent work in QG is thus important because it does not merely suggest that spacetime fails to exist in some sense, it also suggests that there may not even be a C-series. Assuming that the loss of a C-series is the kind of discovery that would lead people to deny that time exists, such a discovery turns out to be compatible with QG. This, in turn, suggests that the loss of the C-series, and thus time in the folk sense, is compatible with what we know from science. We can capture the central idea here as the following argument:

P1   There are live physical theories that eliminate the C-series.

---

[4] See Amelino-Camelia (2002) for an overview.

P2   If T is a live physical theory then the situation described by T is an epistemic possibility.

C   So, the elimination of the C-series is an epistemic possibility.

We call this the quick argument for timelessness. A live physical theory is any physical theory that is currently taken seriously by scientists and isn't considered to be refuted. All confirmed scientific theories are live, but not all live theories enjoy confirmation. If our best, live physical theories treat T as epistemically possible, then a suitably naturalistic outlook demands that we treat T as an epistemic possibility.

Before we take a closer look at some approaches to QG, and thus make a case for P1 in the argument, we need to consider a wrinkle. As discussed in Part One of the book, the vignettes that lead people to deny that time exists all lack a C-series. However, as we also discussed, it is conceivable that individuals were not responding to the lack of a C-series but, instead, to the absence of some other feature from these vignettes. This poses a problem for the conclusion that we want to draw from the quick argument for timelessness, for it suggests that the epistemic possibility that the C-series is missing may not be enough to show that the kinds of discoveries that lead people to deny that time exists are compatible with what we know. For there may be some *other* feature that leads people to make these judgements, and the absence of this feature may be incompatible with what we know about the world, even if the absence of the C-series is not.

There are two things to say here. First, one of the vignettes that led people to deny that time exists—the D-theory vignette—is based on one approach to QG—Barbour's (1994a, 1994b, 1999) Machian approach. Moreover, other approaches to QG paint a picture of reality that is very much like the one captured by the D-theory vignette in many respects. It thus seems quite likely that *whatever* features lead individuals to deny that time exists are missing in at least some approaches to QG, which is all that we really need.

Second, the problem at issue cuts both ways. The challenge from our opponent is that there might be features missing from the vignettes other than the C-series that lead people to deny that time exists. The trouble is that we have no idea what these features might be. If it is not the discovery that the C-series is missing, then we are simply ignorant of what the discoveries might be that are responsible for people's judgements. That makes it very difficult for our opponent to defend the idea that the discoveries at issue are incompatible with what we know about the world.

This line of thought will become important later, and so it is worth pausing a moment to explore the idea. We began this chapter with a challenge: the kind of discoveries that would lead people to reject the existence of time are incompatible with what we know. Now, if we know what such discoveries might be, and in particular that such discoveries involve the absence of a C-series, then the challenge seems compelling. In order to respond to the challenge, we would then need to show that the C-series might not exist, which is what the quick argument from timelessness seeks to establish. If, however, we don't know what discoveries lead people to deny that time exists, then our opponent's challenge does not seem very good. For they don't seem to be in a position to establish that such discoveries are incompatible with our best science, given that they can't even tell us what the discoveries are. It is thus rather easy to establish that the loss of time in the folk sense is compatible with what we know. For if we don't even know what the loss of time involves, we simply can't rule it out scientifically, or otherwise.

We will return to this issue in light of some things we say about approximations to spacetime in Chapters Seven and Eight. For now, we will focus on the C-series and on the quick argument for timelessness. In the next section, we will put some flesh on the bones of the quick argument by taking a brief look at some proposals for a theory of QG.

5.3. Three Approaches to Quantum Gravity

There are a number of approaches to QG available. For present purposes, we will consider just three: Barbour's Machian approach to canonical quantum gravity, loop quantum gravity, and causal set theory. We focus on these three approaches because they appear to be the most devastating when it comes to time.[5] In all three approaches, it does not seem that anything like a C-series remains. Note that in what follows we aim to keep technical details to a minimum. Good technical overviews of each theory exist elsewhere. Our

---

[5] A number of philosophers maintain that string theory is likely to be non-spatial. This is due to strange relations of duality between various different string theories with different dimensionalities. The relations of duality are taken by some as evidence that the spatial aspect of the theory is a mere mathematical artifice. As far as we can tell, there is no similar pressure to believe that time, in the form of a C-series, is missing from string theory. That being said, a notion of space may be needed to formulate a viable C-series. If there is no space, it is unclear whether it is possible to individuate the members of the C-series that are supposed to be temporally ordered. After all, the standard way of doing this is to treat the entities that are ordered by the relevant temporal between-ness relation as three-dimensional spatial configurations. For discussion of the string theory case, see Huggett and Vistarini (2015), Huggett (2017), and Rickles (2011). For an overview, see Le Bihan and Read (2018).

goal is just to give the reader a sense of the metaphysical picture of reality that each theory paints, which can then be used to support the quick argument for timelessness.

5.3.1. Canonical Quantum Gravity

The first approach we will consider is Barbour's version of canonical quantum gravity. Canonical quantum gravity takes the standard quantization techniques used in quantum field theory to quantize other fields (such as the strong force) and applies these techniques to the gravitational field, in order to provide a quantum account of gravity. There are serious technical difficulties with formulating a theory of gravity along these lines, most notably: the so-called problem of time, which turns out to really be a host of different problems, depending on which aspect of time one is interested in. Very roughly, though, the problem is that the most plausible way to overcome the technical issues that arise from the application of standard quantization techniques to the gravitational field involves dispensing with time.

Barbour's approach lies squarely within the canonical tradition. His project has two stages to it, both of which he considers to imply timelessness. First, he provides a Machian formulation of GR. This formulation involves converting GR into its Hamiltonian form. To get a feel for this, imagine a configuration space of three-dimensional points that specify the location of every particle, as well as all of the inter-particle distances between them. The configuration space itself represents all of the states that a physical system moves through as it evolves in time. For now, suppose that the configuration space does not contain every possible state for the system, but only the states it actually occupies. We define a dynamics over the system—roughly a law that describes the way that the system evolves—by identifying the path of least action through the configuration space. The path of least action is the most energy-efficient sequence of states for the system to move through.

Barbour proposes to apply this basic idea to the universe as a whole. We begin with a configuration space for the entire universe, rather than a single physical system. We then move to a relative configuration space by removing information about the orientation of the system, and its centre of mass. We go on to identify a path through the space via a method of best-matching, which is supposed to identify for us the path of least action. The process of best-matching can be understood in the following way. Imagine taking a particular point in the relative configuration space as a starting position. Then take all the other points in the configuration space and compare them to that starting point. The 'next' point is the configuration that features the smallest amount of difference to the starting position. Having defined this point, we then repeat the process until we have formed a trajectory through the space. Because we are constantly matching for the smallest difference, the idea is that the trajectory we map out using the best-matching method will be extremely efficient, and thus will be a path of least action.

We can think of a path through a space of this kind as a law that describes the entire history of a four-dimensional universe: a way that all the particles in the universe evolve through time. This corresponds to a Machian formulation of a theory. Mach was a proponent of a relationalist account of space and time. For Mach, space and time do not exist independently of the physical constituents of the universe, namely: matter. Rather, space and time are defined in terms of the relations *between* the material constituents of nature. The best-matching algorithm over configuration space is a way of defining the temporal evolution of the system in terms of the relations between the particles in the relevant configurations, and so is Machian in this sense.

Barbour takes the Machian approach one step further. Rather than thinking of the points in the relative configuration space of the universe as configurations of particles, he takes them to be three-geometries. What this means, in effect, is that the relative configuration space is now a configuration of three-dimensional Riemannian spaces. The relative configuration space is populated by various states that space itself takes (rather than particles in space). In this sense the account is less Machian than it might otherwise be. These three-dimensional Riemannian spaces are not, themselves, defined in terms of the relations between particles. Rather, the points in the configuration space appear to be substantival in nature.

The Machian dimension of the theory enters the picture in the following way. The three-dimensional points in relative configuration space are converted into a four-dimensional picture via the same process of best-matching. This time, however, we take all the actual configurations of space, and then build a path of least action through the configurations based on constantly matching for the smallest differences in the geometry of Riemannian spaces. The result of the best-matching procedure is a spatiotemporal manifold describable by GR. Rather than time being a relation between bits of matter in this picture, time is a best-matching relation between bits of space. It is not a dimension through which space moves, but more like an algorithm that defines a dynamics. Barbour views the theory as timeless because while one can plot a four-dimensional history through the relative configuration space, he maintains that the configuration space itself is the fundamental arena of reality. The configuration space, however, possesses no temporal or spatio-temporal metric structure.

By formulating GR in Machian terms, against the backdrop of a configuration space, Barbour then proceeds to provide a quantum description of the system. He does this by expanding the configuration space. Rather than thinking of the configuration space as just the actual configurations of Riemannian three-geometries, he proposes to treat the configuration space as the space of all possible configurations. This is advantageous, because the Schrodinger wave equation of a quantum system can be defined over the possible configurations of that system. Usually, the configuration space for the wave equation is limited to just a single system. It is, however, possible to define a wave equation over a relative configuration space of every possible configuration of the entire universe, yielding the Wheeler-DeWitt equation. Thus, having formulated GR in terms of a configuration space, Barbour simply expands it with more possible configurations to house the Wheeler-DeWitt equation, thereby yielding a quantum description of the gravitational system that the configuration space describes.

The Wheeler-DeWitt equation is the central result of the canonical quantum gravity program.[6] The equation, however, has no time parameter, which has led a number of physicists to conclude that time is simply absent at the most basic level of reality. As Lam and Esfeld (2013, p. 287) put the point:

> The Wheeler-DeWitt equation does not contain any explicit time parameter, unlike the Schrodinger equation for instance...this is one aspect of the 'problem of time', which finds its roots in the background independence and diffeomorphism invariance of classical GR already. One way to deal with this 'problem of time' is to accept that it indicates time might not be a fundamental feature of the world.[7]

Barbour thus sees his theory as timeless for two reasons.[8] First, the Wheeler-DeWitt equation appears to provide a description of the dynamics of the entire universe *without* appealing to time at all. For, as it is commonly noted, the Wheeler-DeWitt equation does not contain a time variable. As Barbour puts

---

[6] The Wheeler-DeWitt equation is a field equation that attempts to reconcile, mathematically, the wave equation and the field equations of general relativity. See Dewitt (1967).

[7] For further discussion of the so-called problem of time in the context of canonical quantum gravity, see Anderson (2006, 2009, 2012a, 2012b), Kuchar (1992).

[8] For further discussion, see Baron, Evans, and Miller (2010).

the point, the equation gives us a 'dynamics without dynamics'. The second reason that Barbour takes his approach to be timeless concerns his particular interpretation of the equation. As noted above, Barbour takes the equation to apply to relative configurations of the universe as a whole. The underlying space of the equations, then, has no temporal or spatiotemporal metric structure. If time were to show up in the underlying space that the equation describes, then perhaps one could explain away the apparent timelessness of the equation. But that's precisely what we can't do on Barbour's view.

The difference between Barbour's Machian formulation of GR and the Wheeler-DeWitt description of the dynamics is that while one can plausibly reconstruct a four-dimensional history of the universe from the relative configuration space that underwrites Machian GR, there is little hope of doing this in the case of the Wheeler-DeWitt equation. Reality is not a four-dimensional object, like it is in GR. The entire configuration space in the Wheeler-DeWitt equation is taken to be our reality. While one can plot a path through the space for a system that might be considered a temporal path (this would just be the path of least action through the space for a part of the universe), there is no sense in which the temporal path corresponds to the way that our universe is at a global scale. Any way of constructing a temporal dimension of the space would be arbitrary and would involving applying extra physical constraints to the dynamical description of the world.

We are now in a position to draw two morals from Barbour's view. First, because Barbour appeals to a relative configuration space as the basic arena of reality, his approach to QG eliminates the C-series. There is no sense in which points in the space are temporally ordered, and there is no sense in which they are at some distance from one another. Since both ordering and metricity appear essential to the C-series, the C-series does not exist on Barbour's view. Second, because the basic arena of reality is a relative configuration space, the structure of reality resembles one of the vignettes that we used to probe the folk concept of time in Part One. The D-theory vignette that triggered widespread abandonment of time among the folk is a version of Barbour's configuration space picture. In this vignette, reality is modelled as a set of cards that have internal spatial structure, but no connected temporal structure. This just *is* a simple description of Barbour's configuration space. We should thus expect that if the world turned out to be the way that Barbour imagines, then such a discovery would be the kind of discovery that leads people to deny that time exists.

5.3.2. Loop Quantum Gravity

This brings us to the second approach to quantum gravity: loop quantum gravity or LQG.[9] In some sense, LQG is the inverse of Barbour's project. Whereas Barbour attempts to develop a physical picture that allows him to build up to an account of QG, LQG starts with an account of QG and then builds down to a physical picture of the world. Thus, while Barbour's endpoint is constructing a configuration space that can underwrite the Wheeler-DeWitt equation. Loop quantum gravity theorists start with a solution to the Wheeler-DeWitt equation, and then attempt to produce an interpretation of that solution. Thus, LQG begins with a Hamiltonian formulation of general relativity, and then applies standard quantisation techniques of the kind found in canonical approaches to quantum gravity to provide a quantum account of the gravitational field.

According to the kinematical formulation of LQG, physical reality can be described at the most fundamental level by spin-networks. A spin-network can be represented by a directed graph of interconnected vertices and edges. Each vertex in a spin-network represents a three-dimensional volume of space, and each edge represents a two-dimensional surface. Each spin-network is taken to represent a quantum state of space, and reality itself at the fundamental level is thought to be in a super-position of spin-network states (see Figure 5.1).[10] Spin-networks can be embedded into a bare manifold—roughly, a manifold that has been stripped of its metric structure. Rovelli (2004), however, prefers a non-geometric interpretation of loop quantum gravity, according to which spin-networks are represented in a purely algebraic fashion (as the directed graphs of group SU2). He thus aims to capture spin-networks without appealing to any fundamental geometric structure, and so without embedding them inside a bare manifold.[11]

The kinematical version of LQG is generally supplemented by a dynamics to produce a full-blown approach to QG. Roughly speaking, the dynamics of LQG involve extruding the spin-network structure through a higher dimension, resulting in a 'spin-foam' (see Rovelli and Vidotto 2014). The dynamics generally involves the application of a Hamiltonian operator to the spin-network

---

[9] Our presentation of this approach follows Huggett and Wüthrich 2013 closely. See also Perez (2013), Rovelli (1991, 2004), and Rovelli and Vidotto (2014).

[10] Each edge is labelled by a number that is related to the spin-value from particle physics. Similar networks are used in, for instance, quantum field theory to represent quantum spin-state transitions. The spin-networks of LQG should not be taken to represent particle spin in quite the same way.

[11] For further discussion of the various ways in which one may provide an ontological interpretation of LQG, see Norton (2020).

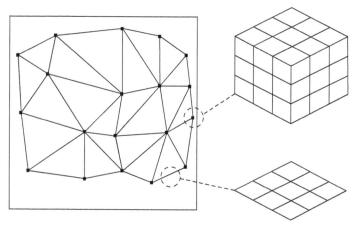

Figure 5.1. LQG.

structure, under which spin-network states can 'persist' or undergo 'splitting' or 'joining' to produce new nodes and edges. We place these notions in scare quotes, because there is little reason to consider the higher dimensional spin-network structure to be a temporal one, and thus it is not really the case that spin-network states persist, split or join in any temporal sense. In this way LQG resembles Barbour's theory, in so far as it features a kind of 'dynamics without dynamics', i.e., the dynamical version of the theory manages to be dynamical without being temporal.

The picture painted by loop quantum gravity appears to be one according to which fundamental reality is constituted by discrete 'chunks' of space that are tied together and that can then undergo various operations of splitting or joining in the dynamics.[12] In this way, the picture resembles a lattice-like structure. Importantly, however, the linking does not introduce any spatio-temporal metric into the picture, nor does it introduce any temporal ordering or metric structure (cf. Huggett and Wüthrich 2013, Wüthrich 2017, 2019). While one may be able to define up a metric by determining a distance function over spin-networks (node x is n nodes away from node y in the directed graph), the metric does not have the right properties to be a temporal or spatiotemporal metric. For one can define a metric of this kind over any directed graph (and over many mathematical objects beside). But simply defining a metric structure of this kind is not enough for have a temporal or spatiotemporal metric structure (consider, for instance, that one can define a

---

[12] Note that there is some disagreement over whether the spin-network structure is discrete or continuous; see Dittrich and Thiemann (2009).

metric over the real numbers, but doing so is not sufficient for the numbers to be temporally related to one another).

Note that there is a sense in which one can treat the edges in a spin-network as being like adjacency relations between chunks of space. After all, the edges are 2D surfaces which themselves connect 3D regions of space. The trouble, however, is that there is no clear way to directly map these adjacency relations to the metric relations needed to produce a spatiotemporal metric (that being said, the mapping may be possible under certain conditions, more on this a bit later). One of the difficulties is that adjacency in a spin-network does not, in general, correspond to anything like spatiotemporal adjacency. Indeed, it is possible to map adjacent nodes in a spin-network to regions of spacetime that are arbitrarily far from one another. As Lam and Wüthrich (2018, p. 48) put the point:

> ... the fundamental adjacency relations represented by the spin network links need not-—and in general do not—correspond to the spatiotemporal, metrical contiguity in the standard GR sense: two connected nodes at the LQG level may correspond to events in GR spacetime that are arbitrarily far away (in the GR metric sense) from one another. In this perspective, the fundamental spin network connectivity cannot be directly interpreted in standard spatiotemporal terms.

Now, Norton (2020) has recently argued that there are many different interpretations of loop quantum gravity available. Some of these are more friendly to the idea that spacetime exists than others. Above we mentioned two interpretations: one according to which the spin-networks are embedded in a bare manifold, and another according to which the spin-networks are represented in purely algebraic terms, and so are not represented as having any geometric or topological properties. For interpretations where the spin-networks are embedded in a bare manifold it is tempting to just call the bare manifold stripped of its metric structure spacetime, and thus maintain that the theory is a spacetime theory after all. Whether one chooses to call loop quantum gravity a spacetime theory is not really an issue we wish to adjudicate here. What matters, for us, is whether there is any metric structure to spin-networks of the kind that might allow us to say that the physical structure being represented has time in it. Since we are working with the assumption that the lack of such structure is sufficient to say that time does not exist, we are inclined to continue treating loop quantum gravity as a timeless theory *even if* one changes the extension of the term 'spacetime' to accommodate a bare manifold, or even a bare manifold with an embedded spin-network (Norton reserves the phrase 'quantum spacetime' for this kind of case).

LQG, then, appears to be a theory that lacks a C-series. Note, however, that there is still plenty of structure in the theory. Space makes a cameo appearance in the nodes and edges of spin-networks. Proponents of LQG also maintain that it is possible to identify derived structure based on the mathematical foundations of the theory, structure that corresponds to something like space-time. Thus, it is thought that spacetime, in some sense, emerges. Exactly what that means, and the extent to which time can be considered genuinely emergent in QG will be considered in greater detail below.

Compared to the configuration space that lies at the heart of Barbour's program, the spin-network structure in LQG is a much more structured entity. On Barbour's picture of reality, the physical arena of the world is a set of isolated spatial points. In LQG, by contrast, the physical arena is a deeply connected entity, one that can be modelled as a graph. Ultimately, however, the two pictures are not that far away from one another. In both cases, there is no temporal ordering or metric structure to be found.

LQG has more structure than the scenario presented in the D-theory vignette discussed in Part One and that led to people giving up the existence of time (in light of the graph theoretic connections in the theory). This is one place, then, where it might be that extra structure left out of the vignette (aside from C-series structure) could be what drives people to respond that time does not exist. We doubt it, though, given that the folk don't generally think in terms of graph theoretic structure, and so it's hard to see how this could make much of a difference to people's views about time.

5.3.3. Causal Set Theory

The third approach we will consider is causal set theory (CST) (see Wüthrich 2012 for an overview). This approach is based on two key ideas. First, there is a widely held expectation that, in quantum gravity, the continuum nature of spacetime will break down at high energy scales, giving way to a discrete structure. Second, there are a number of theorems (such as the one due to Malament (1977)) showing that the metric relations of spacetime can be derived from the causal structure of a spatiotemporal manifold. As Dowker (2014, p. 19) puts the point:

The structure of spacetime that takes center stage in understanding the physics of GR [general relativity] is its *causal structure*. This causal structure is a *partial order* on the points of spacetime. Given two points of spacetime, call them A and B, they will either be ordered or unordered. If they are ordered then one, let us say A without loss of generality, *precedes*—or, is in the past of—the other, B. This ordering is *transitive*: if A precedes B and B precedes C then A precedes C. The order is *partial* because there are pairs of spacetime points such that there is no physical sense in which one of the pair precedes the other, they are simply unordered. This lack of ordering does *not* mean the points in the pair are simultaneous because that would imply they occur "at the same time" and require the existence of a global physical time for them to be simultaneous in … global physical time does not exist in GR.

The partial ordering of points to which Dowker refers corresponds to the light-cone structure of spacetime. Everything that is in the forward light-cone of a point A follows A causally: any such point is one to which causal influence may propagate from A. Everything that is in the backward light-cone of a point A precedes A causally: any such point is one from which causal influence may propagate to A. The forward and backward light-cones of A do not exhaust the points in spacetime: there are points outside of A's light-cone entirely, points that lie in A's absolute elsewhere. These are points that cannot be reached from, and cannot reach, A by causal influence of any kind. There is no fact of the matter as to whether these points precede A or not. Hence the sense in which the ordering of points that corresponds to the causal structure of spacetime is partial and not total.

According to causal set theory, underlying apparently continuous spacetime we find partially ordered causal sets. Each causal set is an ordered pair of $\langle C, \preceq \rangle$. According to Dowker, the elements of causal sets are not three-dimensional volumes of space. Rather, they are supposed to be four-dimensional 'atoms'. Here's Dowker (2014, p. 21):

Let me stress here a crucial point: the elements of the causal set, the discrete spacetime, are atoms of four-dimensional spacetime, *not* atoms of three-dimensional spacetime. An atomic theory of space would run counter to the physics of GR, in which three-dimensional space is not a physically mean-ingful concept. An atom of spacetime is an idealization of a click of the fingers, an explosion of a firecracker, a here and now.

The last part of this passage from Dowker gives us some insight into how 'large' a spacetime atom is, i.e., what, exactly, its four-dimensional extent is supposed to be. Four-dimensional spacetime atoms are supposed to be *idealized* in a certain respect. Elsewhere she clarifies the idea when she is talking about spacetime events. She writes:

> Spacetime is the collection of all idealized *events* where an event is something that happens at a point in space at a moment in time, something like the popping of a champagne cork. These events are 'idealized' because a real champagne cork occupies some spatial volume and the pop lasts for a finite period of time; a spacetime event can be thought of as the limit of smaller and smaller corks shorter and shorter pops.   (Dowker (2006, p. 2)

Although Dowker is here talking of events in spacetime, we take it that her intended interpretation of 'idealized' in the case of spacetime atoms is the same. A spacetime atom is the limit of an event that is extended in time. We can think of these atoms, very roughly, as four-dimensional spacetime atoms with an infinitesimal extent in the time-like axis.

As noted, the second element of the causal set picture is the relation $\leq$ over causal sets. This has the following properties:

(1) If $x \leq y$ and $x \leq z$ then $x \leq z$, $\forall x, y, z \in C$
(2) If $x \leq y$ and $y \leq z$ then $x=y$, $\forall x, y \in C$
(3) $\forall x \in C, x \leq x$
(4) For any pair of fixed elements x and z of C, the set $\{y|x \leq y \leq z\}$ of elements lying between x and z is finite.

The first three properties are transitivity, anti-symmetry, and reflexivity respectively, rendering $\leq$ a partial order. The fourth property rules out continua, by ruling it out that there is an infinite number of elements between any two elements in a causal set ordered by $\leq$. The fourth condition thus forces causal sets to form a discrete structure.

A system of causal sets forms an upward branching tree. These structures can be modelled via Hasse diagrams, which model finite, partially ordered sets. Whereas a family tree grows down, a Hasse diagram grows upwards via the partial order $\leq$ (see Figure 5.2). Causal sets come into existence gradually, through a random process defined by a particular probability distribution. Once they come into existence, they remain in existence as members of a partially ordered causal set. The partial order thus unfolds via a stochastic 'process' called the law of sequential growth. This gives us the dynamics of the theory. Dowker describes this law as follows:

> Each model [of causal set dynamics] is a stochastic process in which a causal set grows by the continual, spontaneous, random accretion of newly born elements. To describe the process requires the introduction of "stages" labelled by the natural numbers. At stage N of the process, there is already in existence a partial causal set of cardinality N, and a new causal set element is born. It chooses, according to the probability distribution that defines the process, which of the already existing causal set elements to be to the future of, thus forming a new partial causal set of cardinality N + 1. This can be thought of as a transition from one causal set to another with one more element, with a certain transition probability. Then at stage N + 1, another causal set element is born—another transition occurs—and so on.
>
> (Dowker 2014, p. 21)

Crucially, the use of natural numbers here to index each stage of the stochastic process should *not* be taken to represent the use of an absolute time variable, such as might be found in Newtonian mechanics, as the 'arena' for this process. Rather, the natural numbers are left un-interpreted, as merely indicating the ordering of the stages themselves.

CST is yet to deliver a full-blown strategy for quantizing the gravitational field. It is, rather, more of a step along the road to a theory of QG or a partial attempt to develop such a theory. Still, let us suppose that CST broadly captures the structure of reality. Does reality so described feature time? One might think so, for three reasons. First, the causal set approach makes indispensable use of causal structure and proposes to take causal structure as the fundamental structuring principle of reality. But one might think that time is necessary for causation. If so, then merely having a causal structure is enough

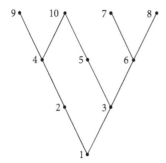

Figure 5.2. Hasse diagram representing a causal set.

to have a temporal structure as well. This line of thought is not compelling, however.

It is not clear that ≤ is a causal relation in anything like the normal sense of causation. The ≤ relation is just a partial order, and any partial order that preserves the discreteness of the causal set lattice would do. It has not been shown that ≤ possesses the features that we normally associate with causation. It has not been shown, for instance, that ≤ supports counterfactual dependence, say, or that it features in causal laws. It is just a piece of mathematical structure and treating it as causation in the normal sense, the sense that presupposes time, is to adopt a metaphysical position that seems to go beyond what the theory demands.

The second reason for thinking that the causal set approach features time concerns the nature of the elements of causal sets. The elements are, after all, supposed to be four-dimensional atoms of spacetime, and so there is a sense in which there is time in the basic posits of the theory. There are two things to say here. First, it's not at all clear how seriously we should take the idea that the atoms are temporal. All the theory really needs, so far as we can tell, is for the atoms to be four-dimensional; it is not obvious that one of those dimensions needs to be a temporal one. Moreover, even if the elements of the causal sets are temporal in some sense, they are not related to one another by a temporal metric or any temporal ordering. Each element of a causal set is more like an isolated, instantaneous world, one that is cut off from any of the others, temporally speaking.

The third reason for thinking that the causal set theory features time relates to the interpretation of the theory itself provided by Dowker, Sorkin, and Rideout. These physicists all interpret the stochastic process in virtue of which causal sets grow (in the manner modelled by a Hasse diagram) as a process embedded in time. Indeed, Dowker goes so far as to describe the stochastic process of causal set growth as a form of temporal becoming, whereby spacetime atoms come into existence in a manner that is analogous to growing block theories of time. The worry here, however, is that this interpretation of the causal set programme appears to require embedding the entire discrete structure within an extra time-dimension, in which the 'growth' of causal sets unfolds. Perhaps this would be plausible if there were temporal ordering relations between the elements of causal sets that could yield metric connections, but there aren't. As Huggett and Wüthrich (2013, p. 278) put the point:

There simply is nothing on the fundamental level corresponding to lengths and durations (or, more generally, to spacetime intervals), and no alternative interpretation of the causal sets in terms of metrical relations is available.

The lack of temporal ordering between the elements of the causal sets (given that ≤ is not a temporal ordering or metric connection) seems to undermine the interpretation that Dowker et al. offer. More generally, it provides evidence that there is no C-series structure in the theory.

Arguably, CST offers a much more structured picture of reality than either LQG or canonical quantum gravity. Still, the theory removes a C-series ordering and that's enough for our purposes. As with LQG, the picture of reality painted diverges from the D-theory vignette that we used to trigger responses about the reality of time. Given this, it is open that there is some feature missing from that vignette but present in CST that would be at odds with the kind of discovery that leads people to deny that time exists. Since we don't know what kind of discovery that might be, however, we are inclined to lean on the response to this issue that we offered above. Namely, that since we don't know what discovery *other* than the absence of a C-series leads people to deny that time exists, we aren't in a position to rule that such a discovery is incompatible with CST.

5.4. Emergence

In addition to canonical quantum gravity, LQG and CST, the loss of time has been indicated for a number of other approaches to QG.[13] As Wüthrich (2019) puts the point, in each case:

> ...the structures postulated by these theories lack several, or most, of those features we would normally attribute to space and time, such as distance, duration, dimensionality, or relative location of objects in space and time.

Exactly how spacetime, and space and time, go missing, depends very much on the approach to QG under consideration. Our goal, however, is not to provide a detailed analysis of the various approaches to QG. Rather, we will simply follow Wüthrich and others in assuming that some approaches to physics are

---

[13] These are by no means the only approaches to QG available, and are not the only approaches that are thought to be timeless.

lacking the kinds of structures flagged in the passage above. What matters to us is whether the loss of space, time, and spacetime in the context of QG, implies that the *C-series* in particular goes missing. The answer appears to be that it does. If Wüthrich is right that spatial and temporal *distance* is generally lost, then the C-series must be missing. After all, a C-series is, at a minimum, an ordering of events in time that gives rise to a metric and, as such, can be used to specify how far apart members of the ordering are from one another. If there is no way to do this for a theory, then there just is no C-series to be found.

Of course, we recognise that Wüthrich's appraisal of the state of the art in QG is not without controversy. Philosophers have argued that, in at least some approaches to QG, time (or something time-like) remains.[14] But we recognise a presumption in favour of Wüthrich's view and so, for the purposes of this book at least, we assume that at least some approaches to QG—approaches like LQG and CST—eliminate a C-series. This, in turn, provides a basis for accepting the quick argument for timelessness. There is, however, an obvious challenge to that argument. While many philosophers agree that space, time, and spacetime are eliminated *in some sense* by some approaches to QG, there is nonetheless a widely held expectation that spacetime will be an emergent phenomenon. The emergence of spacetime, one might argue, implies that spacetime exists. If the existence of spacetime is a necessary feature of QG, however, then the discoveries that lead people to reject the existence of time are incompatible with what we know after all, since as discussed the existence of spacetime is likely sufficient for the existence of a C-series.

A similar move is made for a great many entities that play no part in fundamental physics. Chairs, cats, dogs and people are all missing from our most fundamental physical theories, yet we don't thereby conclude that these entities do not exist. Rather, we accept only that they don't exist *fundamentally*, but of course we allow that they exist at some non-fundamental level. And so it is for spacetime. Despite not being a fundamental component of reality, spacetime nonetheless exists at some non-fundamental level.

Call this: the emergence challenge. The challenge can be set out schematically, in argument form, as follows:

---

[14] For instance, Le Bihan and Linnemann (2019) argue that quasi-space and quasi-time exist in LQG on the grounds that the covariant form of the theory still obeys the Lorentz symmetries. Gryb and Thebault (2016) argue that, in at least some interpretations of CQG, there is a lingering temporal structure to be found.

P1   For any theory of QG in which spacetime does not exist fundamentally, spacetime will emerge.

P2   If spacetime emerges, then spacetime exists.

P3   If spacetime exists then the C-series exists.

C   So, for any theory of QG in which spacetime does not exist fundamentally, the C-series exists.

Note that by 'spacetime' we mean relativistic spacetime, and thus have in mind the *specific* entity posited by the general theory of relativity (GR). Following a number of other philosophers, we take the ontology of GR to consist in at least some entity represented by a smooth manifold equipped with the metric field described by the metric tensor in the Einstein field equations. We focus on this entity because we take it as given that if this thing exists, then so does the C-series, and so the argument is strongest in this form. The thought, then, is that spacetime in this sense emerges, despite not appearing in a fundamental theory of QG. If one has something else in mind by the phrase 'spacetime' then a connection between that thing, whatever it is, and the C-series will need to be argued for. Perhaps this can be done, but it really depends on what else 'spacetime' might be. We will return to say a bit more about this issue in Chapter Six and Chapter Seven. For now, however, we will proceed with the understanding that spacetime is GR spacetime.[15]

The appeal to emergence is common in the philosophical literature on QG. Emergence, however, is a troubling notion. The term 'emergence', as Paul Humphries once quipped, is a phrase that has escaped semantic control: there are a great many things that are called 'emergence' within philosophy, and very little agreement on what does, or does not count, as a case of emergence. That being said there are two broad ways in which the term gets used when discussing the emergence of spacetime.

The first of these is what we will call *theoretical* emergence. This notion of emergence is a relationship between theories; one closely linked to deducibility. As Huggett and Wüthrich (2013, p. 277) put it 'spacetime [is] in some way derived or (to use the term in a very general sense, as physicists do) 'emergent' from the theory.'

According to Crowther (2018), *theoretical emergence* in the context of QG has two important dimensions: *dependence* and *independence*.[16] Crowther

---

[15] Earman (2002) argues that GR is timeless in some sense, and perhaps this implies that there is no C-series if GR is true. However, see Maudlin (2002) for a response.

[16] Crowther's account extends the work of Butterfield and Isham (1999).

characterises the dependence dimension of emergence primarily in terms of a broadly Nagalian conception of reduction (Nagel 1935). Thus, according to Crowther, a theory T2 depends on a theory T1, when T1 is more fundamental than T2, where T1 is more fundamental than T2 when the laws of T2 depend on those of T1. And where the laws of T2 depend on the laws of T1 when, effectively, the laws of T2 reduce to the laws of T1. More carefully, T2 depends on T1 when there is a corrected version of T2, T2* such that the following four conditions hold of T2*:

1. The primitive terms of T2* are associated via bridge laws with the various terms of T1.
2. T2* is derivable from T1 when it is supplemented with the bridge laws specified in 1.
3. T2* corrects T2 in that it makes more accurate predictions than T2 does.
4. T2 is explained by T1 in that T2 and T2* are strongly analogous to one another, and T1 vindicates why T2 works as well as it does in its domain of validity. (Crowther 2018), p. 80)

The independence dimension of theoretical emergence requires there to be a certain degree of autonomy of T2 from T1. Crowther characterises this autonomy in terms of two features. First, novelty. T2 should differ from T1. Moreover, the difference must be substantive in so far as the description of reality provided by T2 cannot be expressed using the theoretical apparatus of T1. Second, robustness. T2 should be invariant under various changes to T1. For a number of different ways of altering the description of the underlying physics provided by T1, T2 remains derivable. In other words, T1 in a certain sense underdetermines the physics described by the more fundamental theory. This first notion of emergence is not, in the first instance at least, supposed to be a metaphysical notion.

The second notion of emergence we will call *metaphysical emergence*. Metaphysical emergence, as we will understand it, is a form of ontological dependence and so, at a bare minimum, x is metaphysically emergent from y when, if x exists, then y exists and if x exists, it exists in virtue of the existence of y. Following Bedau (2008), we take metaphysical emergence to have two further features. These features mirror Crowther's two conditions of theoretical emergence almost exactly, but are to be construed in a more metaphysical spirit. First, when some entity B emerges from some entity A, B is dependent on A. Second, despite the fact that B is dependent on A, B is autonomous from A in a certain respect. For instance, B might be a novel phenomenon in the sense that it possesses properties or features that A does not possess. Or, B might be robust with respect to changes to A, in this sense: there is some class of changes to A that don't result in changes to B, thereby imbuing A and B with different modal properties.

To clarify, when we are talking about metaphysical emergence we are not talking about 'strong emergence', which sometimes goes by the same name or by the name 'ontological emergence'. Strong emergence is typically taken to imply a failure of deducibility. Thus, if y strongly emerges from x, then a description of y cannot be derived from a description of x. Strong emergence, then, implies a failure of theoretical emergence.[17] Now, it may be that space-time and spatiotemporally located entities are strongly emergent from the underlying ontology of a theory of QG.[18] But as we shall see, there are reasons to suppose that this is not so, and so we will assume that strong emergence is off the table. Rather, we will take the ontological emergence of spacetime and of spatiotemporally located entities to be compatible with the deducibility of a description of the emergent phenomena from a description of the base phenomena. This kind of emergence is usually referred to as 'weak emergence'.

In light of our distinction between theoretical emergence and metaphysical emergence, we now have two ways of sharpening up the emergence challenge stated above. The first sharpening treats 'emergence' as theoretical emergence; the second sharpening treats 'emergence' as metaphysical emergence (which may or may not bring with it deducibility). Our goal for the rest of this chapter is to consider the theoretical version of the argument. We will return to the metaphysical version of the argument in Chapter Six. In the reminder of the chapter we argue that theoretical emergence does not undermine the quick argument for timelessness.

5.5. Theoretical Emergence

The theoretical version of the emergence argument can be stated as follows:

P1   For any theory of QG in which spacetime does not exist fundamentally, spacetime is theoretically emergent.

---

[17] See Chalmers (2008) for an overview of the distinction of weak and strong emergence. For further discussion, see Barnes (2012), Humphreys (2016), and Wilson (2016).

[18] For further discussion of emergence in quantum gravity, see Baron (2019), Bain (2013), Chalmers (2021), Crowther (2016), Le Bihan (2018a, 2018b, 2021), Lam and Wüthrich (2018), Lam and Esfeld (2013), and Yates (2021).

P2   If spacetime is theoretically emergent, then spacetime exists.

P3   If spacetime exists then the C-series exists.

C   So, for any theory of QG in which spacetime does not exist fundamentally, the C-series exists.

Let us hold fixed that GR spacetime can be derived from QG—that is, let us grant that spacetime is theoretically emergent. Does the fact that we can derive GR from QG imply that spacetime exists (and thereby undermine the quick argument for timelessness)?

Our answer: there is no simple reason why this should be so. At the most general level, theoretical emergence is compatible with both realist and anti-realist attitudes towards the emergent phenomena. To see this, it is useful to compare two distinct examples of theoretical emergence in science.

First, consider the inter-theoretic relationship between thermodynamics and statistical mechanics. Thermodynamics describes a range of macro-level properties of matter, including work, energy, heat and radiation. Statistical mechanics provides an explanation of these macro-level properties in terms of the micro-level constituents of matter: particles. Thermodynamics appears to be an emergent theory in Crowther's sense. It provides a description of novel phenomena (there is no heat at the micro level), and one that is underdetermined by the underlying physics of statistical mechanics (many different particle systems correspond to the same thermodynamic system). Moreover, thermodynamics is, or at least is thought to be, derivable from the underlying statistical mechanical theory.

The theoretical emergence of thermodynamics is standardly coupled with a realist attitude regarding the novel phenomena that the emergent theory describes. We believe in the existence of heat by virtue of the role that it plays in thermodynamics. The discovery that one can derive thermodynamics from statistical mechanics is not typically accompanied with the pronouncement that we have discovered that heat does not exist (on the grounds that it plays no role in the statistical mechanical description). We do not abandon heat in favour of, say, the mean kinetic energy of particles. Rather, we are inclined to treat the two theories as offering distinct ontological descriptions, which are both apt descriptions of reality albeit at different levels.

But now compare the thermodynamic case with the emergence of Newton's mechanics from Einstein's. Einstein's theory of relativity becomes effectively Newtonian under certain conditions. In particular, for a group of test particles that are moving with a low velocity compared to the speed of light, situated in a weak gravitational field that is unchanging over time, it is possible to use a Newtonian description of the system with minimal error. Indeed, under an appropriate generalisation and geometrical formulation of the Newtonian theory, namely Newton-Cartan theory, it is possible to derive Newton's mechanics from Einstein's.[19] Thus, in this sense Newton's mechanics serves as a mathematical limit for general relativity. Newton-Cartan theory is, however, formulated against the backdrop of a different spatiotemporal structure to Einstein's theory.[20] Whereas Einstein's theory requires a Lorentzian spacetime, Newton-Cartan theory is formulated against a Galilean spacetime.[21] Rather than a single spatiotemporal metric field—of the kind posited by GR—a Galilean spacetime features two, orthogonal degenerate metrics that correspond to space and time respectively.

Newton-Cartan theory and GR thus have different ontologies. GR posits the existence of the spatiotemporal metric field; Newton-Cartan theory posits a structure featuring two orthogonal dimensions: space and time. There is, we maintain, no requirement to believe in the existence of both the metric field of GR and the degenerate metric structures found in Galilean spacetime. On the contrary, the features of the world that the Newton-Cartan theory seems to describe or require are features that we can be anti-realist about, despite the fact that we can derive Newton-Cartan theory from GR.

Indeed, it is presumably *because* we can derive Newton-Cartan theory from GR under certain conditions that it is permissible to eliminate the ontology of the derived theory. When the derivation goes through, we are able to show that all of the empirical predictions of the Newtonian theory are derivable from GR, because the Newtonian theory is derivable. In a certain sense, then, we don't *need* to believe in the existence of anything over and above the metric field of GR in order to accommodate the empirical results of both GR and Newton's mechanics. That's because the metric field of GR effectively behaves as if it were Newtonian within the regime where Newton's mechanics is

---

[19] Named for Cartan (1923), Newton-Cartan theory was subsequently developed by Dautcourt (1964), Havas (1964), Künzle (1976), and Ehlers (1997). For a broad discussion of the Newtonian limit of general relativity in mathematical terms, see Carroll (2003).

[20] A precise account of the similarities and differences between the two varieties of spacetime is not important for present purposes. What matters is that Galilean spacetime displays distinct behaviour to a Lorentzian one. In particular, in a Galilean spacetime there is an absolute time variable; but not so in Lorentzian spacetime. The Newton-Cartan theory is also modally robust compared to Einstein's. There are many various different solutions to the field equations that retain Newton-Cartan theory as a limiting case. Thus, just as thermodynamics is theoretically emergent, so too Newton-Cartan theory appears to be theoretically emergent from Einstein's theory of relativity.

[21] For an account of Galilean spacetime see Ehlers (1973) and also Earman (1989). For discussion of the spacetime structure of Newtonian theory see Knox (2011, 2014), Weatherall (2016), and Saunders (2013).

relevant. That entity, then, can do everything that the ontology of Newton-Cartan theory can do (and more). Unlike the thermodynamic case, there is no need to double our ontological commitments in light of the theoretical emergence of one theory from another.

One might be inclined to push the argument a bit further. In the case of Newton-Cartan theory, it is not just that one can reasonably withhold a strong ontological commitment to the entities posited by that theory; the very fact that one can derive Newton-Cartan theory at all using GR may provide a reason to eliminate the entities posited by the derived theory based on a perceived incompatibility between the two ontologies.

To see the idea, suppose there is a specific region, containing a group of test particles. When the gravitational field is weak and static, we can use Newton-Cartan theory to accurately model the dynamics of those particles. Suppose, however, that for whatever reason the gravity in that region is subsequently increased. Then we can no longer use the Newtonian theory and must revert to GR to give an adequate description of the dynamics. What are we to make of the region at issue, ontologically speaking? Is it the case that the region contains both the metric field of GR and the orthogonal metric structures of the Newtonian theory, one riding atop the other? Or does the metric field come and go, somehow 'transforming' into strictly Newtonian structures as gravity is lowered? Neither option seems very good. The first option poses a threat of incompatibility: the two metric fields are not obviously compatible with one another and so it is not clear that, metaphysically speaking, they can obtain in the same region. The second option requires a new metaphysics of metric transformation that is difficult to understand. Much better, we think, to commit only to the ontology of GR, and have it function in various ways, under various conditions.

To be clear, we are not saying that one *cannot* believe in the existence of Newtonian structures. One can still take on the extra ontological commitments of Newton-Cartan theory if one wishes, assuming that the addition of ontology over and above the metric field of GR can be well-motivated. Nor are we saying that one should not, under any circumstances, believe in the existence of the Newtonian structures. It may be that an argument can be provided for believing in both the metric field of GR *and* the specific posits of Newton-Cartan theory. Our point is just that *mere* derivation does not *force* ontological commitment. It is compatible with the fact that Newton-Cartan theory is derivable from GR that the ontology of the derived theory is eliminated in favour of the ontology of the base theory.

Still, one might remain worried: accepting the ontology of Newton-Cartan theory, one might argue, amounts to nothing more than accepting the existence of certain approximately realised structures. Since the metric field of GR does, in fact, approximate the posits of a Newtonian theory, then we are committed to the ontology of the Newtonian theory after all. We disagree. To see why, it is useful to draw a distinction between two kinds of ontological commitment: strict and loose. A strict ontological commitment is a commitment to the ontology of a theory as described *exactly* by that theory. A loose ontological commitment, by contrast, is a commitment to something that is sufficiently close to whatever is posited by a specific theory.

Our claim is that no *strict* ontological commitments follow from the derivability of Newton-Cartan theory from GR. Which is to say that we are not forced to accept the existence of time and space as specified by the Newtonian theory. Of course, we are forced to accept the existence of something that approximates the Newtonian structure in virtue of the derivation, and so if one wishes to speak loosely then we are forced to hold a commitment to entities posited by that theory in the loose sense. This means that we must accept that the GR metric field possesses certain properties or exhibits certain behaviour under specified conditions. And to be sure there is some measure of ontological commitment to be registered as a result. But nothing is forced on us in the strict sense, at least not from derivation alone.

No doubt there are other examples to be found of the differing attitudes one might take in the face of theoretical emergence, but the two examples discussed are sufficient to show that theoretical emergence on its own does not settle questions about ontology. Whether or not one should believe in the existence of the entities or properties described by the emergent theory is often a further question, one not settled by the fact of theoretical emergence itself. Thus, it may be that the emergence of GR spacetime in QG is just like the Newton-Cartan case. In which case, the description of spacetime in GR may simply give way to a description of a different, non-spatiotemporal structure in QG, one that is capable of approximating GR spacetime under certain conditions. On the other hand, the theoretical emergence of GR from QG may be more like the thermodynamic situation, insofar as a commitment to the metric field of GR plus the ontology of QG is required.

5.6. Conclusion

There is obviously much more to say about theoretical emergence in science and in QG. We have by no means provided a comprehensive discussion of the topic. We have said enough, however, to see where, in a general sense, an emergence challenge to the quick argument for timelessness based on theoretical emergence goes wrong. The theoretical emergence of GR from QG underdetermines any ontological conclusions. To make a determination of what exists, we must look beyond the inter-theoretic relationships between GR and QG and consider the metaphysical relationship between the structures posited by GR and the structures posited by QG. This leaves us, then, with metaphysical emergence to consider. Here too we believe the quick argument for timelessness survives the emergence challenge. We argue for this in the next chapter, in which we consider a range of different options for cashing out the metaphysical emergence of GR spacetime from QG. We argue, in each case, that the metaphysical emergence of spacetime from something non-spatiotemporal faces substantial problems.

6

Metaphysical Emergence

6.1. Introduction

In this chapter, we focus on the metaphysical emergence of GR spacetime from a more fundamental, non-spatiotemporal ontology. As discussed, this kind of emergence presents a way to formulate what we have called the 'emergence challenge'. The emergence challenge can be stated in terms of metaphysical emergence as follows:

P1   For any theory of QG in which spacetime does not exist fundamentally, spacetime is metaphysically emergent.

P2   If spacetime is metaphysically emergent, then spacetime exists.

P3   If spacetime exists then the C-series exists.

C   So, for any theory of QG in which spacetime does not exist fundamentally, the C-series exists.

In our discussion of theoretical emergence, we argued against a parallel argument by challenging P2 of that argument. There, we argued that the theoretical emergence of spacetime does not imply the existence of spacetime (though it is compatible with it). As we see it, there is little hope of offering a similar response to P2 of the argument above. If, as noted, metaphysical emergence involves ontological dependence, then so long as a commitment to the base entity can be secured, then the existence of the emergent entity quickly follows. Since, in this case, the base entity is a posit of QG, then an ontological commitment to such an entity would be sufficient for the existence of GR spacetime, if GR spacetime is indeed metaphysically emergent. And, in that case, the quick argument for timelessness will fail.

Our aim, instead, is to challenge the claim that GR spacetime is metaphysically emergent, and to thus argue against P1 in the metaphysical form of the emergence challenge. Note that for this argument to pose a problem for the quick argument for timelessness, the first premise must be understood in a particularly strong fashion. The quick argument for timelessness only seeks

Out of Time: A Philosophical Study of Timelessness. Sam Baron, Kristie Miller, and Jonathan Tallant, Oxford University Press.
© Sam Baron, Kristie Miller, and Jonathan Tallant 2022. DOI: 10.1093/oso/9780192864888.003.0006

to show that it is epistemically possible for there to be no C-series. The metaphysical emergence argument needs to close this possibility, and to do that the first premise must show that spacetime *must* emerge in QG.

We challenge this claim by considering a number of different ways of making the notion of metaphysical emergence more precise. Each option involves specifying emergence in terms of a particular metaphysical relation. The trouble, however, is that the options are all spatial, temporal or spatio-temporal notions (or tacitly rely on the same), and so without spacetime it is unclear that we can make sense of metaphysical emergence in each case. Because standard ways of understanding metaphysical emergence appear to break down for spacetime, it is not so clear that spacetime must be emergent. At best, it seems that it *might* be emergent, but equally it might not be. That's because the difficulties we outline here may signal a general breakdown of metaphysical emergence as applied to the emergence of spacetime.

One final point before we get started. One might worry that, in what follows, we are considering metaphysical emergence in the context of only 'fringe' theories of QG. Our argument that metaphysical emergence faces problems would thus not be particularly interesting, since it would only apply to theories of QG that aren't taken very seriously anyway. Worse, it wouldn't apply to the theories considered in Chapter Five that are used to establish the quick argument for timelessness in the first place (assuming these are not fringe theories). In fact, however, our goal is to show that metaphysical emergence in general, regardless of the underlying theory of QG, faces problems when it is used to characterise the relationship between spatiotemporal and non-spatiotemporal structures. Thus, the argument does not target only fringe approaches to QG but is fully general.

6.2. Relations of Metaphysical Emergence

To quickly rehearse two points from the previous chapter, metaphysical emergence—like theoretical emergence—is taken to have at least two features. First, when some entity B emerges from some entity A, B is dependent on A. Second, even though B is dependent on A, B is autonomous from A in a certain respect. For instance, B might be a novel phenomenon or B might be robust with respect to changes in A. The dependence dimension is here understood as a relation R of something akin to ontological dependence. But which relation is it? There appear to be three broad options.

First, R might be a mereological relation of composition. Thus, to say that spacetime depends on a more fundamental ontology specified by a theory of QG, is to make a claim about parthood. A view along these lines has been defended by Le Bihan (2018a, 2018b), who maintains that spacetime has a range of non-spatiotemporal parts.

Second, R may be a relation of material constitution. Material constitution is familiar from discussions of the statue and the clay. The statue, it is often supposed, is constituted by the clay from which it is sculpted. So too it might be the case that spacetime is constituted by a range of more fundamental entities.

Third, R could be functional realisation. If R is a relation of functional realisation, then the relationship between GR spacetime and the more fundamental ontology of QG is analogous to the relationship between mental states and neural states. Thus, just as mental states like pain exist in virtue of the fact that neural states come to play a certain functional role, so too may we say that GR spacetime exists because the underlying ontology of QG plays a specific functional role.

In addition to composition, constitution, and realization, one might be tempted to appeal to two other relations often used in metaphysics to cash out ontological dependence. The first of these is the (now somewhat retro-chic) relation of supervenience. The second is the more contemporary notion of grounding. We don't consider these to be viable options for understanding the metaphysical emergence of spacetime. We'll explain why, starting with supervenience, before turning our attention back to composition, constitution, and functional realisation.

Underlying the concern that we will spell out for both supervenience and grounding in 6.1.2 and 6.2.2 respectively is a methodological issue surrounding the metaphysical emergence of spacetime, one that is worth bringing to the fore. In giving an account of the relation in virtue of which spacetime emerges, it is not enough to simply posit some relation or other and be done with it. Such an approach leaves the matter far too unconstrained. What we seek, and what we think ought to be found, is an account of the metaphysical emergence of spacetime that is explanatory, in this sense: it yields some understanding of how it is that spacetime might emerge from a more fundamental non-spatiotemporal reality.

The reason for this requirement is that there appears to be a conceptual gulf between a picture of reality as fundamentally non-spatiotemporal, and a picture of reality as spatiotemporal. Le Bihan draws attention to this conceptual divide by drawing an analogy to the hard problem of consciousness. It is difficult to see how mental phenomena might arise from purely physical phenomena, like neural states. For some, this is due to a gap between our concepts of mental phenomena and our concepts of physical phenomena. To then simply say that mental states arise from physical states in virtue of some primitive grounding relation does not seem enlightening, precisely because it does not yield much by way of understanding that might help us to overcome the apparent conceptual divide between the mental and the physical. The issue in the case of spacetime is very similar. It is difficult to see how spatiotemporal phenomena could arise from non-spatiotemporal phenomena. Thus, an adequate account of the metaphysical emergence of spacetime should yield some insight into how spacetime exists and should not merely encode the fact that it exists (if it does).

To make the request vivid, consider the case involving composition and how that might serve to give us a sense of what the emergence of space-time might consist in. We have at least some conceptual grip on what it means for one thing to be composed of others. If we could somehow carry this over into the case of spacetime being made from non-spatiotemporal parts, then we'd have a similar sense of how the metaphysical emergence of spacetime is supposed to work.

As we will argue below, this is precisely what one cannot do: our conceptual grip on parthood hits a wall when it is applied beyond a spatiotemporal context. Our point, however, is that at first glance one's pre-theoretic grasp on what composition is enables one to see, if only in a dim way, how extending mereology to the emergence of spacetime might be meaningful. What we doubt in the case of supervenience and grounding is that these relations can provide us with the same dim grip on how spacetime could be emergent.

6.2.1. Supervenience

Lam and Esfeld (2013) briefly consider supervenience as a way to understand the dependence of GR spacetime on a more fundamental structure. They dismiss it, arguing that:

> Supervenience implies covariation in the following sense: any variation in type B-properties necessarily involves variation in type A-properties. However, there is no account available of how a variation in spatio-temporal

properties could involve a variation in the properties of a more fundamental entity that is not spatiotemporal; again, one encounters the difficulty how to conceive change without space and time.   (Lam and Esfeld 2013, p. 292)

While we agree that supervenience should be struck off, we don't think that it should be excluded because the relation itself is essentially spatiotemporal in nature.

Lam and Esfeld's (2013) reason for thinking that supervenience is a spatio-temporal relation appeals to change. The idea, roughly, is that supervenience presupposes change, and change requires space and time.

In making this argument, Lam and Esfeld appear to confuse two types of 'change'. On the one hand, there is change over time. They are correct that it is difficult to conceive of change in this sense in the absence of space and time. But that is not the kind of 'change' that is implicated in the notion of variation that underpins supervenience. The relevant notion of variation is modal. When B supervenes on A, there cannot be a 'change' in the B-properties without a change in the A-properties. What this means is that any modal variation in the B-properties implies some modal variation in the A-properties. In short, if the B- properties had been different, then the A-properties would have been different as well. This modal variation does not require there to be any actual (temporal) change in the B or A properties, it is to be understood entirely in terms of what would happen were we to make a hypothetical alteration in some other possibility.

We agree with the conclusion of Lam and Esfeld's argument. Our concern is that supervenience relations are not dependence relations. There can, after all, be patterns of supervenience without any dependence. For example, every mathematical fact trivially supervenes on every other mathematical fact, assuming that mathematical facts are necessary. But it is not the case that every mathematical fact depends on every other mathematical fact. Rather, there are very specific patterns of dependence within mathematics that quite clearly come apart from supervenience facts. All that a supervenience claim does is attribute a pattern of modal co-variation to the phenomena of interest, it does nothing to say why that pattern exists or how it arises. To explain that—in this case, to explain how we get a temporal structure from a non-temporal underlying structure—we need some account of the depend-ence relation between the supervenient entity and its subvenient base. As above, we are seeking some kind of explanation. Supervenience gives us nothing.

6.2.2. Grounding

A more promising option is the much-discussed grounding relation.[1] Grounding, we suppose here, is an asymmetric, transitive, irreflexive, non-monotonic dependence relation. Grounding thus does not suffer from the flaws inherent to supervenience when it comes to cashing out dependence: dependence is baked in.

We can differentiate between two approaches to grounding. The first of these is to treat grounding as a way of unifying a broad group of relations. On this conception, grounding is the generic name that we give to any relation that might be an ontological dependence relation, and that has the right formal properties (namely, those just specified). Candidates to be grounding relations in this sense are just the relations specified above: composition, constitution and functional realization.[2] As such, if we understand grounding in this first way, then it does not represent a distinct option to the three already outlined.

There is, however, a second understanding of grounding available in the literature. According to this understanding, grounding is not a generic name for a class of relations. It is, rather, a unique, primitive metaphysical dependence relation over and above the relations already specified.[3] If we understand grounding in this way—if we eschew an interpretation of grounding as at least one of composition, constitution and functional realization—can we make sense of spacetime emergence?

We think not. Our complaint, in a nutshell, is that appealing to a primitive ontological dependence relation to account for the metaphysical emergence of spacetime amounts to *naming* the problem rather than *solving* it. We already know that there should be some relation of ontological dependence between spacetime and some non-spatiotemporal structure. If we say that the relation is one of grounding, and all we can say about it is that it's an ontological dependence relation, then we seem to have just recast the original question of exactly what the ontological dependence relation might be between spacetime and a more fundamental structure.

Perhaps this is unfair: in taking grounding as primitive, we have achieved a bit more than merely giving a name to the very relation we are seeking to understand. We have also said that the relation is fundamental, and primitive,

---

[1] For an overview of grounding in the sense we have in mind here, see Schaffer (2009b).

[2] Some include the determinate/determinable relation among the list of grounding relations. We do not consider this option here, since we don't really see how to understand the relationship between a fundamental, non-spatiotemporal structure and spacetime in determinate/determinable terms.

[3] For discussion of this way of thinking about grounding, see Wilson (2014) and Koslicki (2015).

and therefore that there's nothing more to be said about it. That's something. But we don't think it's enough. We have no problem with allowing the use of primitives within metaphysics. But primitives must pay their way: the extent to which we can reasonably take a certain aspect of the metaphysics as primitive is to be judged in terms of the kind of explanatory work we can do when the primitive is inducted into whatever metaphysical system we might be working with. Our concern, here, is that appealing to a primitive grounding relation as the basis for emergent spacetime does no explanatory work. What we don't get is any understanding of how it is that spacetime emerges, metaphysically, from some more fundamental structure that is not spatiotemporal. If the relationship is a primitive one, then we are effectively forbidden from asking any further questions about the metaphysics of dependence in this case. We must take whatever understanding can be gleaned from the mere fact that there is ontological dependence. But it is difficult to see how much by way of understanding can be gathered from this fact.

The problem with grounding in the primitive sense just described, then, is that we don't really have much of a pre-theoretic grasp on the notion. So, we don't really learn anything about how spacetime emerges from a fundamentally non-spatiotemporal reality when we apply that relation. As discussed above, we think that an account of metaphysical emergence should, to a certain degree, yield some understanding of how it is that spacetime emerges from a more fundamental, non-spatiotemporal reality.[4]

One might disagree. Granted grounding *on its own*, when construed as a primitive relation does not yield much by way of understanding when applied to the case of spacetime emergence. However, it does not follow that a grounding-based picture will be un-explanatory. For instance, one might try to supplement grounding with a story about how some fundamental, non-spatiotemporal structure grounds a non-fundamental, spatiotemporal one. Thus, grounding might still be the right story for the nature of the dependence; it is just that we need to spell out how the grounding works in order to get some explanatory traction on spacetime emergence.

We think that's right: if one supplements grounding with a story about how the grounding of spacetime works, then a grounding-based picture can be

---

[4] One might worry that the constraint is an instance of metaphysics overreaching: it is not the metaphysician's job to explain how it is that spacetime emerges. This is, rather, a job for physics. To be clear, our claim is not that the complete answer should be given by metaphysics. Of course, whatever answer one gives must be appropriately informed by the physics of the situation. But when the physics remains silent, or otherwise fails to fully determine the nature of the metaphysical emergence, as seems to be the case in QG, then it is up to the metaphysician to pick up some of the slack.

explanatory. Our objection to the use of grounding as a primitive only really applies if one refuses to provide any further story about how the grounding works.

But what else can one say about how grounding works in the case of spacetime emergence? The difficulty is that the options for spelling this out appear to be broadly the options we consider in this chapter: parthood, composition, and functional realisation. To illustrate: let us agree that grounding is a primitive relation. We might still suppose that the spacetime that is grounded by the timeless fundament is thereby a part of it. The grounding is what gives rise to this mereological structure, and it is through the mereological relations between fundamental and derivative that we can begin to see *how* the grounding relation does its work: it generates time-ful parts of a timeless base.

And, this is so even if grounding is not used as a basis for unifying these relations; for spelling out how grounding works generally proceeds via some such relation. Thus, whatever explanatory power comes from a grounding-based account, appears to stem from these other relations. And since we do not think that these relations are in good standing as relations between the fundamental and derivative, as we explain in the sections that follow this one, we do not think that these relations can work in tandem with grounding, here. Ultimately, then we don't see what explanatory gains there are to be had by appealing to grounding itself. The value seems to lie in how grounding works, and thus in the physical-cum-metaphysical story that one can tell about this.

6.3. Parts of Spacetime

We turn now to a discussion of the three candidate relations for metaphysical emergence introduced above, starting with composition. If the metaphysical emergence of spacetime is understood in terms of composition, then the way that spacetime depends on a range of more fundamental entities is by having those entities as parts. The dependence of spacetime on something more fundamental is thus analogous to a range of more familiar cases of dependence within science. Specifically, the dependence of spacetime is very similar to the dependence of macroscopic objects on their microscopic constituents.

Although initially promising, there are three related problems with appealing to the part-whole relation as the basis for the metaphysical emergence of spacetime, at least in so far as the part-whole relation is standardly understood in metaphysics.

6.3.1. On the Nature of Location

The first problem relates to location. There is thought to be a tight relationship between the location of the parts of an object and the location of the whole. This locative relationship shows up in at least three places in the literature on mereology. First, it is one aspect of what Sider (2007) calls the 'intimacy' of parthood. Sider (2007, p. 25) describes this intimacy partly in terms of the following inheritance principle:[5]

> **Inheritance of Location (IL):** if x is part of y, then y is located wherever x is located.

The inheritance of location is, according to Sider, one of the conceptually significant aspects of parthood. As he puts the point:

> The locations of a thing's parts are automatically reflected in the thing's location—more evidence of the intimate nature of the part-whole connection. Parthood is alone in this respect; my location is not tied to the locations of my relative, things I own, things I am near, and so on...our actual usage of 'located' reflects the peculiar intimacy of parthood; we choose a meaning that matches my boundaries with my parts' locations...Everyone accepts the inheritance principles. If they are true, then the part-whole connection is a uniquely intimate one. (Sider 2007, p. 25)

Second, the relationship between location and parthood also shows up in the notion of mereological harmony. Mereological harmony is the idea that the mereological relations between entities are reflected in the mereological relations between the locations of those entities. Mereological harmony is thought by some (e.g., Schaffer (2009a)) to be a necessary truth for objects. Saucedo (2011) formalises mereological harmony as the conjunction of eight separate principles. The fifth principle imputes a relationship between the sub-region that an object is located at, and the sub-region(s) that any part(s) of that object are located at:

---

[5] According to some (e.g., Simon (1987)), the connection between parthood and location is motivated via an extensional conception of parthood. To be clear, though, we are not presuming that parthood is extensional (i.e., that if x and y share all of the same parts then x = y). Whether or not parthood is extensional is orthogonal to the connection to location identified by Sider. The idea, rather, is just that the inheritance of location is part of our pre-theoretic conception of parthood.

H5 for any x and any y, x is a part of y iff x's location is a subregion of y's location.

This fifth harmony principle resembles Markosian's (2014) sub-region theory of parthood, which is the third place that the relationship between location and parthood appears:

Sub-Region Theory of Parthood (STP): for any x and for any y, x is a part of y iff the region occupied by x is a subregion of the region occupied by y.

According to Markosian (2014), STP is highly intuitive. As he puts the point:

I would say that of all the main answers to all the various questions concerning the mereology of physical objects, STP is probably the most intuitive. Only a complete mereological radical would deny that occupying a subregion of the region occupied by an object is a *necessary* condition for being a part of that object.   (p. 5)[6]

IL, H5 and STP all seem to be tracking the same basic idea: it is part of our concept of parthood that the location of the whole is reflected in the location of its parts. Thus, if the whole is located in a certain place, then the parts are (or should be) located there too.[7]

If the concept of parthood requires even a fairly minimal connection between the location of the whole and the location of the parts (such as the necessary connection specified above) then it is difficult to make sense of the idea that spacetime or spatiotemporally located entities are composed of entities that are not spatiotemporal. Indeed, the very idea straightforwardly falsifies all three principles, as we now show.

First, consider the inheritance of location. Suppose that there is some entity A that is entirely composed of entities $B_1 \ldots B_n$ that are not spatiotemporally

---

[6] A similar point is made by Braddon-Mitchell and Miller (2006, pp. 223–4), who write in a different context that:

It is at least *necessary* that a proper spatial part is an object that occupies a region of space that is a sub-region occupied by the whole. This minimal necessary condition presupposes very little.

[7] Note that the relationship between location and parthood does not foreclose the possibility of wholes or parts that lack a location entirely. So, for instance, one could still say that a mathematical object, like a pure set, has its members as parts. That's because abstract objects lack a location, and both the set and the members are abstract objects. What the concept of parthood requires is the transmission of location from (at least) wholes to parts, when the wholes are located—this is Braddon-Mitchell and Miller's minimal necessary condition—and (perhaps) from parts to wholes when the parts are located—as enshrined in the biconditional in H5 and STP. Abstract objects satisfy this transmission principle trivially. Problems arise when one relatum of the composition relation is located and the other is not.

located. Suppose, however, that A has a spatiotemporal location. Assuming that the location of the whole is inherited by the location of the parts, then the B_n should be located wherever A is. So, B_n should be located in spacetime. But *ex hypothesi* B_n are not spatiotemporally located. We have a contradiction.

Second, consider H5. Suppose that A is located at a region of spacetime R. By H5, B_n ought to be located at a sub-region of R. But B_n are not located at any sub-region of R, because B_n are not located in spacetime. We have another contradiction.

Third, STP: A occupies a region R, which is a region of spacetime. So, by STP, each of B_n ought to occupy a sub-region of R. But B_n are not spatiotemporally located, and so they do not occupy any spatiotemporal region. *A fortiori* they do not occupy any sub-region of R.

We think that there is scope for our opponent to respond, but we don't think that the responses succeed. For instance, one might see the problem, but maintain that whether the three principles considered above are falsified depends a bit on what A is. Suppose that A is just a piece of matter, swirling around in spacetime. Then the arguments look good: spatiotemporally located pieces of matter cannot be composed of non-spatiotemporally located entities without violating the three principles outlined above. But that just means we shouldn't suppose that spatiotemporally located entities are composed of non-spatiotemporally located parts. Instead, we should suppose that only spacetime is composed of non-spatiotemporally located parts. If A is *spacetime itself*, then the arguments sketched above are no good, one might contend. For, goes the thought, spacetime is not itself spatiotemporally located. We can therefore just suppose that spatiotemporally located entities depend, for their existence, on spacetime first and foremost, and thus on some group of non-spatiotemporal entities at best indirectly. If we assume that the relevant notion of dependence throughout is composition, there is thus no reason to suppose that spatiotemporally located entities are directly composed of non-spatiotemporally located entities: they are, at worst, indirectly composed of them.

But indirect composition is bad enough. The location of a spatiotemporally located entity will 'flow down' to the non-spatiotemporally located entities that lie at the foundations of the mereological chain. To see this, suppose that A is composed of some part of spacetime P, which itself is composed of some non-spatiotemporal entity E. Then by IL, P will be located wherever A is. But then by IL again, E will be located wherever both P and A are, thereby situating E within spacetime once more.

Our opponent could try to reason as follows: the way that spatiotemporally located entities depend upon spacetime is not via composition. There appear to be two further options. First, the 'dependence' might not be dependence at all, it might be identity. But then each spatiotemporally located entity is identical to some part of spacetime. If that part of spacetime is composed of non-spatiotemporally located entities, then the location of the spatiotemporally located entity will flow down to the fundamental ontology once again, via IL. Second, the spatiotemporally located entity might be constituted by some region of spacetime.[8]

Alternatively, one might argue that spatiotemporal entities don't depend on spacetime at all. Rather, spatiotemporal entities are merely located in spacetime, without being dependent on it. Thus, there is no need for a constitution, composition, or identity relation between spatiotemporal entities and spacetime. We just need the location relation. Given this, there is little reason to suppose that spatiotemporal entities will be somehow indirectly composed of non-spatiotemporal entities in virtue of being located within spacetime.

The appeal to location to understand the relationship between spatiotemporal entities and spacetime seems to be the best way forward. Thus, perhaps indeed we can safely ignore such entities and focus just on spacetime itself. Even so, the threat posed by the inheritance of location remains, for there is a good argument for the view that spacetime *is* spatiotemporally located. The argument goes as follows. First, every open region of spacetime is a part of the entire spatiotemporal manifold. Second, every open region of spacetime is also located in spacetime. If, however, every open region of spacetime is spatiotemporally located, and these regions compose spacetime, then spacetime should also be located by the inheritance of location (this time, however, from parts to wholes).

Now, one could deny that spatiotemporal regions are located in spacetime. But this promises to violate IL again. For suppose that there's some spatiotemporal region R and R is not located in spacetime. Then for any entity that is located at R, that entity would not be located in spacetime either (on pain of violating IL). For instance, suppose that Sara is located at R. If R is not located in spacetime, then neither is Sara. But then, where is she? Much better to suppose that R is located in spacetime, and that Sara inherits her location in spacetime from the region she occupies.[9]

---

[8] As we shall see in the next section, however, constitution presupposes composition and so this won't help either.

[9] One might worry that if spatiotemporal regions are located, then any non-empty region of spacetime will result in an unattractive case of co-location. For instance, suppose that France is located at a spatiotemporal region R. It is plausible to suppose that if R is located in spacetime, then it is located at itself. But France just is another spatiotemporal region R*. It follows, then, that R and R* are

We can imagine a further response—though again we don't find it compelling. The response runs as follows. Isn't it odd to say that spacetime is located? Doesn't that require embedding spacetime in a larger manifold and so on ad infinitum? After all, for an object to be located in a spacetime requires the existence of a spacetime. Thus, if we are committed to there being a spacetime that itself has a location, does that not commit us to the existence of a further spacetime in which the first spacetime can be located?

Not at all. Spacetime is trivially located at itself. This is the same for all locations. Where is France? Trivially, France is located in France; Italy is located in Italy and China is located in China.[10] Indeed, according to the standard picture of how location works—outlined by Casati and Varsi (1999)—conditional reflexivity is an axiom of the fundamental location relation (see below).

Conditional Reflexivity: $\forall x \forall y [L(x,y) \rightarrow L(y,y)]$

Conditional reflexivity implies that if anything is located in spacetime, then spacetime is located at itself. If that's right, however, then given what location is, spacetime can't fail to be spatiotemporally located. So, even if the only object that is made up of non-spatiotemporally located parts is the manifold, that is enough to falsify the three locative principles discussed above (IL, H5 and STP). If Sider is right about the conceptual significance of these principles for parthood, it follows that there is a problem with applying the standard concept of parthood to the case of non-fundamental spacetime.

Of course, we recognise that the idea that spacetime is located remains a controversial claim. Even if it is not located, however, we think that a problem remains with the inheritance of location. For whatever else one thinks, it seems clear that spacetime regions are parts of spacetime. We can thus imagine two ways for spacetime to be composed of non-spatiotemporal parts. First, spatiotemporal regions are composed of non-spatiotemporal parts, and those parts compose spacetime. This violates the inheritance of location, because spatiotemporal regions are spatiotemporally located (they occupy a specific position in the manifold), and their parts are not. Second, spacetime is composed of non-spatiotemporal parts directly, not via regions, but it also

---

co-located, despite being distinct regions. This, we might think, is unattractive: spatiotemporal regions can't be strongly co-located at one another. But France is a country. The country occupies a region, namely the region R, but it is not a distinct region. So there's no threat of co-location for regions here.

[10] We concede that, due to a quirk of naming conventions in colloquial English, some china is not located in China.

has spatiotemporal regions as parts. Since regions are spatiotemporally located, then either spacetime is not located and its regions are, which violates the inheritance of location, or spacetime is located and its non-spatiotemporal parts are not, which also violates the inheritance of location. Either way, then, a mereological approach to spacetime emergence can't be squared with locative inheritance.[11]

6.3.2. On Arrangement and Individuation

The next problem facing any proposal to treat the emergence of spacetime in terms of composition is like the problems already discussed in so far as it signals a degree of conceptual breakdown. Consider that two very different objects can be made of the same parts arranged differently. We can, for instance, arrange the internal workings of a clock to produce a timepiece or we can arrange the very same internal workings to produce a steampunk sculpture. In general, composition is not simply a matter of an object being composed of some parts. The object must be composed of certain parts *arranged in the right way*. The standard notion of an 'arrangement' that we apply to our understanding of composition and parthood is a spatiotemporal one. We individuate different arrangements based on the spatiotemporal relations between the parts. This is an important dimension of the difference between the clock and the sculpture: the parts are arranged in the two cases in quite different ways in space and time.[12]

This is particularly so in the context of using composition as a basis for explaining how low-level phenomena might 'give rise' to high level phenomena. If we say that this 'giving rise to' is a matter of mereological composition, then it is natural to compare the QG case with other, similar uses of composition in a physical context toward a similar end. So, consider a table. Tables don't appear in fundamental physics. But that's fine, because we have a physical story about how tables arise from micro phenomena. This story, however, relies crucially on the notion of an arrangement. A table is composed of molecules that are arranged in space and time in a particular fashion. Change the arrangement

---

[11] See Baron (2019, 2021) for an overview of several problems of location that arise when using parthood to understand spacetime emergence.

[12] Arrangement is, of course, not sufficient for individuation. Often, we need to add further conditions to individuate in a more fine-grained way, but at the very least spatiotemporal arrangement is a key component of how we think about composition.

enough and you remove the table from existence (e.g., by blowing it up). The point, then, is that insofar as we understand how to apply mereological notions in a physically meaningful way, we make indispensable use of a notion of spatiotemporal arrangement to explain how the parts 'come together' to produce the whole. The trouble, then, is that the usual notion of 'arrangement' that facilitates the application of mereological notions cannot be used in a context in which the parts themselves are not spatiotemporally located. If we have no spatiotemporal locations at the fundamental level, then we cannot spatiotemporally arrange entities at the fundamental level in such a way that they 'come together' to produce a spatiotemporal whole.

The third problem facing any proposal to treat the emergence of spacetime in terms of composition is like the arrangement problem but applies to the individuation of parts rather than wholes. In order to say that an object is composed of multiple parts, we need a way to individuate the parts from one another. Our usual way of individuating parts makes use of space and time to at least some extent. For example, the table appears to be made of (at least) the following parts: the top, and the four legs. It appears to be at least a necessary condition for differentiating the four legs from one another that they are located in distinct regions of space.[13] If there are no spatiotemporal regions at the fundamental level, it's unclear how we can individuate the locationless parts such that they can come together to give rise to a spatio-temporal whole.

Composition, then, does not seem like a viable way to understand the metaphysical emergence of spacetime from a more fundamental ontology (though we will return to say a bit more about this later). The way that we usually work with the concept to individuate parts and wholes is not applicable outside of a spatiotemporal context, and a core aspect of the parthood concept—its connection to location—is lost. As such, if we are to adopt a position according to which we have metaphysical emergence of spacetime from a timeless base, we will need to understand emergence differently.

---

[13] Perhaps this could be denied. One might argue that *co-location* is possible for parts, and thus there can be two distinct parts of an object that are located at the same spatiotemporal location, and that differ in terms of their other properties. It is difficult to make sense of this for material objects, however. For two distinct material objects that are parts of some third object to be co-located, the two parts would either need to materially interpenetrate one another, or we would need to allow that both are composed of the very same parts (and so do not interpenetrate) but are yet distinct. This is a view that some philosophers take, of course. But it faces the difficult issue of how the resulting objects, which share all of the same parts, are distinct: that is, what grounds their distinctness given that they are composed of the same parts.

6.4. Constitution

This brings us to the second of the three candidate relations for understanding the metaphysical emergence of spacetime introduced above: material constitution. Constitution is the relation that we invoke when we say that one thing is made from another. So, return to our lump, and the statue made from it. When the clay is formed into a statue, many philosophers think both that the statue and lump are distinct, and that the clay *constitutes* the statue. Nonetheless, the lump and the statue appear to have quite different properties. The lump can survive being flattened, for instance, but the statue cannot. Similarly, the statue possesses aesthetic qualities that the lump of clay lacks, and so on.

In the QG case, then, we could say that spacetime is constituted by whatever the fundamental posit or posits might be. This sounds initially promising because, as in the statue and lump case, we can attribute different properties to spacetime compared to the fundamental posits of a theory of QG. So, we can say that spacetime has a particular metric structure that underwrites spatiotemporal location while, at the same time, denying that this is true of the fundamental posits that constitute spacetime.

The trouble, however, is that the standard notion of constitution presupposes composition (cf. Wasserman (2004, p. 694)). If the statue is constituted by the clay, then the statue and the clay share the same parts. Once the clay has been formed into the statue, there is no part of the clay that is not also a part of the statue. If spacetime and the fundamental posits of a theory of QG share the same parts, however, then the difficulties facing the use of composition discussed above arise here too.

In response, one might argue that the part-sharing requirement, as stated, is too strong. Consider a lump of clay that is spread across a table so that, at one end, the clay is formed into a statue of a cat and, at the other end, the clay is formed into a statue of a dog. Clearly, both the dog and the cat statues are constituted by the clay. But the statues do not share the same parts with the clay. The cat statue does not have the same parts in common with the dog statue, despite the two statues being made from the same lump of clay. If total part sharing were a requirement of constitution, then the cat and dog would share the same parts as each other.

But even if we weaken the part sharing requirement so that it gets the case of the cat statue and the dog statue right, we still have a problem. Suppose we make the requirement simply that there must be *some* part sharing in cases of constitution, rather than *total* part sharing. Thus, we can say that the cat statue shares some of the parts of the clay lump, and we can say the same of the dog statue. Since the parts that the cat statue shares with the clay and the parts that the dog statute shares don't overlap (or, at least, don't wholly overlap) we can avoid the counterintuitive consequences of total part sharing noted above. The trouble, however, is that in the spacetime case we must avoid even the smallest amount of part sharing. If the fundamental entities that constitute spacetime share any spacetime point or region as a part with spacetime itself, then it will be hard to resist the view that the fundamentals are spatiotemporal in some sense.[14]

Perhaps, then, we can simply drop the part-sharing requirement altogether. The part-sharing requirement is usually introduced to avoid certain kinds of counterexamples arising for constitution, and so at first glance this might not seem like a very promising line to pursue. However, the usual motivation for the part-sharing requirement doesn't seem to apply to the constitution of spacetime from a non-spatiotemporal ontology. The central motivation for part-sharing stems from the fact that, in the statue and lump case, the statue and the lump are co-located objects. The lump is located in all and only the same places that the statue is located. Recall, however, that the statue and the lump are supposed to be distinct objects. If the statue and the lump are distinct objects, and they are both physical objects, then if they occupy the same location and they don't share parts, then they must interpenetrate one another in a bizarre way. In the QG case, however, the threat of interpenetration never arises. That's because spacetime and the fundamental entities from which spacetime is constituted are never co-located, since the fundamental entities lack a spatiotemporal location. It is also not clear that spacetime is the right kind of entity for material interpenetration anyway.

Even if we drop the part-sharing requirement, however, and sever the direct connection to composition, constitution still seems to share a feature in common with composition that cannot be quite so easily removed. Constitution and composition both seem to imply the inheritance of location

---

[14] In response, one might appeal to Rudder-Baker's (2000) conception of constitution which drops the part-sharing requirement altogether. However, while Rudder-Baker's view drops the part-sharing requirement, it builds in a coincidence requirement. In particular, Rudder-Baker's view requires that the constituted entity and its constituent are spatially coincident. This can't be a constraint on the constitution of spacetime, since the only way that spacetime and some underlying entity could be spatially coincident is if they are both located in spacetime (given that spacetime is spatiotemporally located, and so the underlying entity would need to share that location).

in some form. In the mereological case, wholes are located wherever their parts are located. In the constitution case, a similar principle seems quite plausible, namely:

Constitutional Locative Inheritance (CLI): if x constitutes y, and y is located at L, then x is at least partially located at L as well.

So, for instance, consider once again the lump of clay that has been made into a cat statue and a dog statue. The cat statue is constituted by the clay and is located at one end of the table. And so, the location of the clay that the cat statue is constituted from should at least overlap with the cat statue's location, and indeed it does. The same goes for the dog statue. CLI thus poses the same problem in the case of spacetime constitution as does the putative relationship between composition and location. If CLI is true, then spacetime cannot be constituted by a non-spatiotemporal ontology, because any such constitution would force the non-spatiotemporal ontology to be spatiotemporally located.

In response, one might argue that CLI is nothing more than the part-sharing requirement in disguise. The reason why constituted objects must share a location with their constituents is simply that the constituted object must share parts with its constituents, and parthood involves the transmission of location from the whole to the parts. So, it is because the cat statue is composed of certain parts that it shares with the clay that the clay must be located where the cat statue is located.

CLI, however, seems plausible even in cases where part-sharing doesn't occur. Consider, for instance, an object that is mereologically simple: it has no proper parts. A number of philosophers have argued that mereologically simple objects are at least possible, if not actual. Now, consider one such object. It is a common view about such objects that although they are mereologically simple, they can nevertheless be constituted by matter. Suppose that is right, and now imagine a sphere that has no proper parts but is nonetheless constituted by a material substance. It still seems right to say, of this sphere, that if the mereologically simple object has a location, then the matter that constitutes it shares that location (in this case the two locations wholly overlap).

So CLI doesn't clearly hang on part-sharing. But what reason do we have to believe that principle? The lack of any counterexamples to CLI is a good start. As far as we can tell, there are no known cases of constitution that violate CLI, aside from the proposed violation of the principle that the constitution of spacetime from non-spatiotemporal entities might imply. Another point in favour of CLI is that it is difficult to imagine the principle failing to be true. When we imagine an entity, and then imagine the constituents of that entity, it is difficult to see how the constituents could fail to be located wherever the constituted entity is located, at least so long as we are using our standard concept of constitution as a guide.

The question, then, is whether we can remove CLI from our conception of constitution and be left with something that we understand. We are even more sceptical that this can be done in the case of constitution than in the case of composition. At least with regard to composition, we have the axioms of mereology available that can give us some grip on parthood once it has been shorn clean of any locational inheritance principles.[15] Our conceptual understanding of constitution is all but exhausted by our familiarity with it in everyday cases, all of which seem to involve both location sharing and part sharing of some kind. Once we remove these aspects from the concept of constitution, it is unclear what is left to structure the remaining notion, and to guide its application to metaphysics.

6.5. Mereology Revisited

In sum, composition and constitution as standardly understood appear to be conceptually wedded to spatiotemporal notions. The standard way of understanding these two relations therefore makes them ill-suited to underwrite the dependence of spacetime on a non-spatiotemporal structure. Is there a way to reimagine these relations to make them fit for service outside of a spatiotemporal context? We will focus on composition because, as discussed, it is likely that composition underwrites constitution.

Recall the three problems facing the use of composition: the location problem (parts inherit locations from wholes), the individuation problem (parts are spatiotemporally individuated) and the arrangement problem (wholes are individuated by spatiotemporal arrangements of their parts). Note that both the arrangement problem and the individuation problem arise because of the way that both notions are conceptually coupled to

---

[15] Of course, we maintain that this still leaves composition conceptually impoverished when it comes to applying that notion to physical reality. But the point is that constitution doesn't even have an axiomatic system to help give us some conceptual grip on it.

location. One way to avoid these problems, then, is to decouple arrangement and individuation from location. Paul (2002) defends a notion of 'qualitative' or 'logical' parthood that is conceptually decoupled in this way. Qualitative parthood is a mereological relation between properties rather than objects. Qualitative parthood appears to make use of a non-locative conception of both arrangement and individuation for parts. If, as Paul has argued, qualitative parthood is a genuine parthood concept, then we appear to have a notion of parthood available for the case of non-fundamental spacetime, one that does not require spacetime for the individuation or arrangement of parts.

Paul's idea is that instead of individuating parts physically, in terms of their spatiotemporal locations, we can instead individuate parts conceptually by 'carving them in our minds', as it were. So, for example, we can take the property of being a round square and see—quite plainly—that this property has the properties of being round and being a square as parts. We can also see that these two properties are *different* from one another, and thus we can make sense of the idea that the property of being a round square is composed of two *distinct* parts. The individuation, however, is detached from any notion of location. Plausibly, properties are located in virtue of their instances being located. A round square, however, is an impossible property and so not physically instantiated anywhere. So, the only connection that these properties have to location cannot be implicated in the notion of individuation being employed.

By appealing to qualitative parthood, we can also make sense of arrangements without thinking in spatiotemporal terms. For instance, compare the properties of being a giant cat and being a cat giant. Both properties have, as parts, the two properties 'being a cat' and 'being a giant'. But they are clearly different properties: while every giant cat is a cat giant, not every cat giant is a giant cat (being giant for a cat might still be quite small and so not qualify a cat for being a cat giant). The difference between the two properties at issue is a matter of how their parts are arranged. But the arrangement itself is not a spatiotemporal arrangement, or even a locative arrangement in any other sense. They are *conceptual* arrangements of parts. Of course, unlike the round square case, these two properties can be physically realised and thus their instances can have locations. But the locations themselves don't do any work in specifying the different arrangements of the properties. And, besides, it is straightforward to formulate an analogous case using impossible properties. It may be then that a better picture of the composition of spacetime from parts that are not spatiotemporally located involves Paul's notion of qualitative parthood.[16] However, we think that there are three challenges facing this Paul-inspired picture of spacetime composition.[17]

First, in addition to spatiotemporal relations, we also need the *relata* in those relations: spacetime points. Spacetime points, however, don't appear to be properties, and so it is not clear that we can apply the notion of qualitative parthood to spacetime points in order to explain how these entities are composed of more fundamental, non-spatiotemporal ones.

In order to address this issue, one could combine Paul's (2002) account of qualitative parthood with her bundle theory of objects (see Paul 2012).[18] On this view, objects are mereological sums of properties, which themselves are mereological sums of other properties. A spacetime point, on such a view, could be construed as a mereological sum of, at least, spacetime relations plus other properties (perhaps properties that contribute to the overall topology of spacetime). This helps to address the problem, because all we need to do is compose spacetime relations from more fundamental properties. Once the spatiotemporal relations have been composed, we can then bundle them in order to produce spacetime points. On such a view, being a relatum in a spatiotemporal relation is partly a matter of being composed by spatiotemporal relations. Spatiotemporal relations compose their own relata. This would solve the first problem.

However, adopting a bundle theory of objects in conjunction with Paul's notion of qualitative parthood gives rise to a second difficulty. On the view we are considering, spatiotemporal relations are composed of more fundamental properties that are then possessed by entities that are more fundamental than spacetime. However, it seems plausible that if an entity E has properties P and Q that compose some further property, R, then E has R as well.

For example, consider the property of being stealthy. Let us suppose that the property of being stealthy is composed of two more fundamental properties: being quiet and being agile. Suppose that Sara has the two basic properties that

[16] At first glance, however, it is not obvious how to apply qualitative parthood to the case of spacetime. Paul's notion allows us to make sense of properties having parts. We are, however, concerned with a particular entity—spacetime. But perhaps we can make some progress on this matter by focusing in on spatiotemporal relations. Relations, we might suppose, can be built from other non-spatial and non-temporal properties. These properties, in turn, can be possessed by more fundamental, non-spatiotemporal entities and, in this way, the spacetime relations can be built out of something more fundamental.

[17] A zeroth challenge for this view is in fact that of explaining what the non-spatial and non-temporal properties might be that compose spatiotemporal relations. Let us suppose, however, that this challenge can be met via an appropriate specification of the underlying ontology (for example, the underlying ontology of causal sets or spin-networks may possess topological properties that can be used to build spatiotemporal relations).

[18] An account of spacetime composition along these lines is defended by Le Bihan (2018a, 2018b).

compose the property of being stealthy and that she has them in the right arrangement for stealth. Then it seems to follow that she has the property of being stealthy as well. Indeed, it is difficult to see how Sara could have the two more fundamental properties, how these two properties could compose the property of being stealthy, and yet how Sara could fail to be stealthy. But this is what we need in the spacetime case. We need fundamental entities like spin-networks or causal sets to possess the properties that compose spacetime relations, without themselves standing in the relevant relations (otherwise, we don't have the emergence of the spatiotemporal from the non-spatiotemporal—the world is in fact spatiotemporal all the way down). But if they have the more fundamental properties, then it seems plausible that these fundamental entities stand in the relevant relations that the underlying properties compose. Just as Sara is stealthy because she has the property parts that compose the property of being stealthy, so too is, say, a spin-network spatiotemporal because it has the property parts that compose spatiotemporal relations.

In general, then, we need a way to prevent mereological sums of properties from being possessed by the low-level ontology that possesses the parts of those properties. This may require reimagining property possession as well as both the ontology of objects and the concept of parthood that underwrites the composition of spacetime. We are not entirely sure how this might go. But even if this second challenge can be met, a third problem remains.

Consider a spacetime region R. Presumably, R will be composed of spacetime points. On the view we are considering, spacetime points are themselves composed of spacetime relations, and these relations are composed of further, non-spatiotemporal properties. Thus, spacetime regions are, indirectly at least, composed of the non-spatiotemporal properties that make up spacetime points. Now, let us assume, as before, that each spacetime region has a spatiotemporal location (which, on this view, it has in virtue of being composed of various spatiotemporal relations). The non-spatiotemporal properties that compose spatiotemporal relations presumably don't have a spatiotemporal location. On this view, then, the whole has a location—namely a spatiotemporal one—that the parts—the components of spacetime relations—lack. Thus, the Paul-inspired view we are considering still manages to violate the inheritance of location from parts to wholes because IL is falsified. Spatiotemporal regions are not located wherever (some) of their non-spatiotemporal parts are located.

What we would need, to avoid this, is a notion of composition according to which an object that is spatiotemporally located can be made entirely of parts that have no spatiotemporal location. The resulting conception severs the connection between location and parthood entirely. And whilst Paul's notion of qualitative parthood severs the connection between location and notions of arrangement and individuation for parts and wholes, the kind of disconnect we need is one where the parts are not located anywhere in spacetime, whereas the whole is. Nothing that Paul says about qualitative composition suggests it can allow for such a thing. It thus remains difficult to see how parts and wholes could come apart, locatively speaking. This is just to reiterate the conceptual point that Sider makes: our ordinary conception of parthood has built into it the inheritance of location between part and whole, such that wherever the whole is located the parts are located there as well.

The question, then, is what, if anything, survives once we have stripped the parthood concept of this remaining connection to location? Is the resulting conception a genuine notion of parthood? Well, if one thinks that all there is to the parthood concept is the axioms of mereology, then there may well be a very thin notion of parthood that can play the required role. It is not clear, however, that the resulting notion would be a candidate to be a dependence relation. Part of the difficulty is that the non-axiomatic aspects of the concept guide our application of parthood to physical situations, thereby helping us to understand the relationship between physical objects in mereological terms. If all we have to go on are the axioms, it is not clear that the application of the concept can yield much by way of understanding, and nor is it clear how we can appropriately apply the concept in physical situations.

One potential way forward is to modify the locative inheritance principle to preserve a connection between parthood and location. Recall the principle:

> **Inheritance of Location (IL):** if x is part of y, then y is located wherever x is located.

Instead of IL, one might provide a *conditional* inheritance principle. The idea being that when the parts and the whole are located, then location is inherited from parts to wholes or vice versa, but when either the parts are not located, or the whole is not located, then location is not inherited. This conditional principle can be stated as follows:

> **Weak Inheritance of Location (WIL):** if x and y are both located and x is part of y, then y is located wherever x is located.

WIL allows for some failures of locative inheritance. Moreover, one might think that WIL is independently plausible. For what WIL allows, but IL forbids, is the possession of concrete parts by abstract objects or vice versa. For, presumably, abstract objects have no location. Given this, the antecedent condition in WIL is not satisfied, and so the conditional is trivially true for such cases of composition.

For WIL to be applicable to the case of spacetime emergence, however, it must be the case that the non-spatiotemporal entities that compose spacetime lack locations altogether (not just spatiotemporal locations). For if the fundamental entities that compose spacetime are located in some non-spatiotemporal sense, then either that notion of location will be transmitted up to spatiotemporal wholes, making them non-spatiotemporally located as well *or* the spatiotemporal location of the wholes will be transmitted down to the locations of the parts, making them spatiotemporally located after all.

It is worth reflecting on what the physics would need to be like in order for CIL to provide a viable basis for spacetime emergence. The physics would need to be such that the most basic constituents of reality are not located in any sense, rather than merely failing to be spatiotemporally located. This is rather like saying that the basic components of reality are much more like abstract objects than they are like the concrete entities that physics generally deals with. But, so far as we can tell, there is little reason coming from the physical side of things to suppose that the basic elements of reality are not located at all. Appealing to CIL to achieve spacetime emergence may therefore require reading a metaphysical picture back into the physics that is not warranted (the same point applies to a conditional version of constitutional locative inheritance).[19]

A world in which spacetime emerges in line with CIL also appears quite radical in a metaphysical sense. For it is usually thought that entities with no location at all—like abstract entities—can't do any causal work. If we give up the idea that the fundamental elements of reality are located, then it also seems that we must give up on the idea that there is any causation at the fundamental level. Fundamental reality starts to look at lot more like a realm of mathematical objects, than it does a realm of physical entities, interacting. Of course, we can't rule out that the world is like this. Perhaps indeed the fundamental ontology of the world is a plurality of non-located, causally inert entities. Or perhaps there is a way to have causation without location. There is clearly more work to be done

---

[19] One might respond: but location just means spatiotemporal location. So, if the fundamental elements of reality are not spatiotemporally located then they are not located. We don't see any reason to believe that location is defined in purely spatiotemporal terms, however. Indeed, as we shall discuss in more detail in Chapter Seven, location seems to be a much broader notion than mere spatiotemporal location.

here. And so we leave an opening for philosophers like Le Bihan and Paul to develop an approach to spacetime emergence based on CIL.

6.6. Spacetime Functionalism

This brings us to the third of our three candidate relations for cashing out the metaphysical emergence of spacetime: functional realisation. The basic idea here is that spacetime emerges because it is functionally realised by a range of more fundamental, non-spatiotemporal entities posited by a theory of QG. Call a view along these lines: spacetime functionalism.

Two distinct versions of spacetime functionalism can be found in the literature. The first of these is defended by Knox (2019). Knox's spacetime functionalism is presented as an instance of conceptual analysis. What Knox aims to do is analyse a particular concept of spacetime, as it plays a role within a certain community; the community of physicists. Very roughly, on Knox's view, spacetime is a specific piece of theoretical structure: it is whatever it is, within a theory, that allows one to define a structure of inertial frames which, in turn, can be used to provide the simplest statement of the laws of a given physical theory. Knox's inertial frame functionalism is analogous to analytic functionalism in the philosophy of mind, which seeks to analyse mental concepts in a similar manner. The central difference, of course, is that analytic functionalism aims to analyse our folk mental concepts, whereas Knox's spacetime functionalism focuses on a technical concept within physics. Knox's analysis helps us to understand why it is that the particular entity posited by GR and the entity posited by Newton-Cartan theory, say, can both be considered to fall under the same broad spatiotemporal conception.

Lam and Wüthrich's (2018, 2021) spacetime functionalism, by contrast, does not seek to analyse the concept of spacetime in a general sense. Rather, Lam and Wüthrich are interested in spacetime as it appears in GR. They thus seek to functionally analyse this entity, with the goal of identifying something in a more fundamental theory of QG that can play the spacetime role so specified. Lam and Wüthrich's spacetime functionalism is closer to psycho-functionalism in the philosophy of mind, which seeks to analyse specific mental entities as specified by a particular scientific theory of the mind. These entities can then be linked to those identified in a more fundamental theory.

The two versions of spacetime functionalism are compatible. One can treat Knox's spacetime functionalism as a way of determining the ontological commitments of GR. Indeed, this is precisely how Knox views her functionalism: as a path toward identifying scientific ontology. What exists, according to GR, is just whatever plays the spacetime role in that theory. Having determined what it is that GR posits, one can then apply Lam and Wüthrich's functional analysis to that entity, with the goal of identifying a realiser in a more fundamental theory. So, according to Knox, what plays the spacetime role in GR is the metric field. If that's right, then what GR commits one to is a real field corresponding to the metric tensor: that's what GR spacetime *is*. The question, then, posed by Lam and Wüthrich's functionalism is whether there is anything that can functionally realise the metric field of GR.[20]

In what follows we will focus exclusively on Lam and Wüthrich's functionalism. We do this because, as already discussed, it is the emergence of the specific entity posited by GR that we are interested in. According to Lam and Wüthrich, a functionalist approach to spacetime of this kind proceeds in two stages, which they summarise as follows:

(FR-1) The higher-level properties or entities, which are the target of the reduction, are 'functionalized', that is, they are given a functional definition in terms of their causal or functional role.

(FR-2) An explanation is provided of how the lower-level properties or entities can fill this functional role.   (Lam and Wüthrich 2018, p. 7)

Exactly what the target of (FR-1) might be depends upon what lies within the ontology of GR. Lam and Wüthrich focus primarily on the metric field of GR. (FR-1) thus encodes a requirement to provide a functional characterisation of that entity. A presupposition of the broad spacetime functionalist project is that such an analysis can be provided.[21]

---

[20] This way of viewing the relationship between Knox's functionalism and Lam and Wüthrich's is not quite right. That's because Lam and Wüthrich seem to disagree with Knox about the ontology of GR. This could either be because they disagree with her about the spacetime concept (though agree with her functional analysis) or because they agree with her about the spacetime concept, but disagree with her about what realises it in GR.

[21] Note that this is a non-trivial assumption. Consider the analogous assumption in the philosophy of mind: that mental states can be characterised, in full, in terms of their functional roles. There are powerful arguments that suggest that this can't be done. In particular, there exist conceivability arguments—such as the zombie argument—that purport to show how a functionalist approach to mental states leaves out the phenomenal or 'felt' character of experience. According to Lam and Wüthrich, it is unlikely that any such arguments will succeed in the case of spacetime functionalism. This is because, there is nothing analogous to the subjective quality of mental states that might be left over after a functional specification has been conducted. Without something analogous to qualia, it is unclear how conceivability arguments could gain any purchase.

A functional analysis of the metric field will identify a range of properties or relations that characterise that field.[22] Anything that has the relevant properties, or stands in the relevant relations, thus manages to realise the metric field of GR. The standard way of developing this idea is to produce a Ramsey sentence for the entity at issue (see Lewis (1970,1972) and Ramsey (1931)). A Ramsey sentence is produced by, first, specifying a range of properties or relations that the entity possesses. We then remove any reference to the specific entity that has these properties, within the theory we are considering. So, in the case of GR, we start by fully specifying all of the properties of the metric field as it appears in GR: $P_1 \ldots P_n$. We then provide a statement of the metric field, call it 'a' in GR as the entity that possesses these properties:

$$(P_1 a \wedge P_2 a \wedge P_3 a \ldots \wedge P_n a)$$

To produce the Ramsey sentence, we existentially generalize the above sentence, thereby removing any reference to the metric field of GR, yielding:

$$\exists x (P_1 x \wedge P_2 x \wedge P_3 x \ldots \wedge P_n x)$$

Having specified the role, we can then look to a more fundamental theory to see if there is anything that satisfies the relevant Ramsey sentence. If there is, then something within that theory plays the GR spacetime role. If not, then not.

Lam and Wüthrich argue that in some theories of QG—such as LQG—it is plausible to suppose that the underlying ontology of the theory is capable of playing the GR spacetime role. The basic demonstration that Lam and Wüthrich provide is mathematical in nature. As discussed in Chapter Five, the posits of LQG are spin-networks, which can be represented by directed graphs. At low energies, the volume and area operators over spin-networks achieve numerical agreement with the metric structure of spacetime—what proponents of LQG call a 'weave state'. Lam and Wüthrich interpret this numerical agreement in a functionalist spirit. They suggest that the numerical agreement of the mathematics of weave states with the metric expression of

---

[22] A key point to note is that the functional role for spacetime should not be construed in terms of a *causal* role. After all, if we characterise the functional role for spacetime causally, then whatever plays the relevant functional role must stand in the causal relations that have been specified. However, the fundamental ontology cannot obviously enter into causal relations without being situated in space and time. Fortunately, there is no reason why functional roles have to be specified causally. We can, instead, simply define a functional role in terms of a network of relations where if something stands in the relevant relations, then it is capable of performing a certain function or playing a certain theoretical role. This is broadly the analysis of functionalism that Polger (2007, p. 251), for instance, recommends.

GR provides evidence that spin-networks have the right properties to be capable of playing the GR spacetime role.

Lam and Wüthrich go on to note that spacetime is multiply realisable at the level of spin-networks. There are many different ways of configuring spin-networks that result in a spatiotemporal structure. The fact that many different spin-networks can realise the same spatiotemporal structure makes a functionalist interpretation of spacetime with respect to LQG quite attractive. Multiple realisability is, after all, one of the primary motivations for functionalist approaches more generally (this is certainly the case in the philosophy of mind).[23]

Let us grant Lam and Wüthrich's claim that something plays the GR spacetime role in theories of QG in general, and in LQG in particular. Even if this is granted, it is unclear that the resulting picture is one according to which spacetime is a metaphysically emergent phenomenon. Thus, it is unclear that functionalism of the kind that Lam and Wüthrich propose is a viable way to understand the metaphysical emergence of spacetime.

Here is the source of our concern. All forms of functionalism can be separated into two kinds: role functionalism and realiser functionalism. In order to grasp the difference between these two approaches to functionalism, it is useful to consider the philosophy of mind case once more. Consider, for example, a pain. Suppose that we adopt a functionalist approach to pain, according to which pain is characterised in terms of a particular functional role, a role that is played, in humans, by neural states. Now consider the further question: what, according to functionalism, is pain? Realiser functionalists and role functionalists give different answers to this question. According to realiser functionalism, pain is identified with whatever it is that plays the pain role. Thus, a pain just is the neural state that plays the pain role. By contrast, according to role functionalism, a pain is identified with a higher order property; 'being in pain', which is the property of being in some state or other that plays the pain role.

Either spacetime functionalism is a form of realiser functionalism, or it is a form of role functionalism. Either way, there is a problem. Suppose, first, that spacetime functionalism is a version of realiser functionalism. Then, on this view, spacetime is identified with whatever it is that plays the spacetime role.

---

[23] Of course, as Lam and Wüthrich admit, demonstrating that spin-networks can play the spacetime localising role is just the first step in a broader functionalist project whereby it is shown that weave states are capable of realising the remaining aspects of spacetime. But let us suppose that this can be done, and thus that spacetime can be functionally realised by spin-networks or, indeed, by some other ontology posited by a theory of QG.

So, for instance, in the context of LQG, this would involve identifying space-time with the weave state of a spin-network. The trouble is that this would imply that the underlying ontology that realises spacetime is, itself, spatiotemporal. Weave states *just are* spacetime. Instead of a picture where spacetime depends for its existence on some more fundamental, non-spatiotemporal ontology, we end up saying that the more fundamental ontology is spatiotemporal after all. The whole point, however, was to try and characterise the dependence of spacetime on the ontology posited by a theory of QG that is non-spatiotemporal in nature.

A similar difficulty arises if we adopt a role functionalist interpretation. According to a role functionalist approach to spacetime functionalism, spacetime is identified with a particular property—being spatiotemporal, say—which we can imagine is the higher order property of being in a state that plays the spacetime role. So, for example, consider again the LQG case. Now take the underlying spin-network. The fundamental spin-network, at low energies at least, is in a state that plays the spacetime role; namely, a weave-state. The spin-network thus possesses the property of being spatio-temporal. So, the spin-network is spacetime since it has been identified with the relevant property. But, as before, that won't do. The spin-network is not supposed to be spatiotemporal or have any spatiotemporal properties. As with realiser spacetime functionalism, the functionalist picture appears to involve a kind of flattening whereby spacetime is injected into the funda-mental ontology.

The problem, in a nutshell, is that if something plays the spacetime role in a theory of QG, then it is just not the case that spacetime does not exist fundamentally. Rather, spacetime is a part of the ontology of our most basic theory of physics after all. But this is contrary to the way that physicists themselves present these theories (as theories without spacetime) and to the way that philosophers like Wüthrich interpret these theories—as being ones that are not spatiotemporal. If, however, these theories really are ones that lack GR spacetime, then there cannot be anything that realises the GR spacetime role in these theories. But then we cannot use Lam and Wüthrich's functional analysis as a way of explaining how it is that a non-fundamental spatiotem-poral ontology arises from a fundamental, spatiotemporal one.

There is an obvious reply to this line of argument. Let us grant the functional reduction, and accept that there is, after all, something in a theory of QG that just is GR spacetime. Still, one might argue that it does not follow from this that spacetime exists at the most fundamental level. For the theory may yet specify a range of more fundamental entities that somehow collectively 'make up' whatever it is that plays the GR spacetime role. So, for instance, in the case of LQG, what plays the spacetime role is supposed to be a weave state. A weave state, however, is a complex entity, made up of many entities, each of which corresponds to a node or a link in the directed graph that represents the weave state as a whole. These more basic entities, one might argue, *are not* spatiotemporal, even though they come together to form something that is spatiotemporal.

The difficulty with this style of response is that some account must then be given of how it is that the fundamental non-spatiotemporal entities posited by a theory of QG come together to 'make up' a weave state or similar, which itself is a GR spacetime. The problem is that in order to specify the sense in which the components of, say, a weave state come together to make up that state, it seems we are forced back to using either composition or constitution, since these are the relations by which components make up wholes. But we cannot appeal to such relations in virtue of their conceptual link to spatio-temporal notions and, at any rate, if we could, then the functionalist approach would be otiose, since we could then just specify the metaphysical emergence of spacetime directly, in terms of mereological or constitutive relations.

At this point, one may worry that our argument over generalizes. Consider a standard case of functional realization: the realization of mental states by neural states. It looks like a parallel argument will work here. At the neural level, there are no mental properties. Suppose, then, that mental states are realized by neural states. Then either some neural state just is a mental state, and so neural states have mental properties after all (realizer functionalism), or some neural state has the higher order property of realizing a certain mental state, which just is a mental property, and so the neural state has mental properties after all (role functionalism). Either way, it appears that our func-tionalist account has forced mental properties down into the underlying neural ontology where they don't belong. Since functionalism about the mind is clearly a viable view, one might think, this kind of argument must fail. But if it fails here, then it fails in the spacetime case as well.

There is, however, a difference between the mental case and the spacetime case. In the mental state case we can say that while some neural states are, indeed, identical to mental states, nonetheless the neural level is free of mental properties. That's because the neural level is the level of individual neurons, or groups of neurons that are, only sometimes, mental states, when they reach a sufficient level of complexity. There is thus scope to say that individual neurons, or even groups of neurons lack mental properties, despite certain neural states possessing them. So, the functionalist picture does not force mental properties down onto the neural level after all.

What this response relies on is the claim that individual neurons come together to make up neural states, but that when taken on their own, they lack any mental properties. It is, in other words, the distinction between something like part and whole that allows us to block the imposition of mental properties onto the neural level. We can say that neurons arranged into a certain whole realise a mental state, and thus that the neural whole has mental properties, but the parts of the whole don't have mental properties. This is in virtue of the more general fact that parts need not share the same properties as wholes.

As discussed, however, this kind of response appears unavailable in the case of spacetime functionalism. The trouble is that we can't obviously draw a part/whole distinction of the right kind. We can't clearly say that there are entities in the fundamental ontology that come together to form a whole, and the whole possesses spatiotemporal properties by virtue of realizing spacetime, whereas the parts of the whole do not. We can't obviously say this because it relies on the notion that spacetime has non-spatiotemporal parts.

A parallel version of our argument about mental states fails precisely because we can help ourselves to mereological resources to block it. The argument in the case of spacetime functionalism, by contrast, works because we can't make the same appeal to mereology.

One might respond that we don't need mereology, even in the mental state case. We have assumed that mental states are realized by neural states, where neural states are mereologically composed of neurons. One could argue, however, that there's no need for neural states. Rather, we can just say that mental states are functionally realized by pluralities; pluralities of neurons. A plurality is nothing more than a group of neurons, of course, and thus requires no mereological machinery. A similar view is available in the case of spacetime functionalism. Rather than saying that spacetime is realized by a particular state—a weave state, say—that is mereologically composed of more basic components—like nodes and edges or smaller spin-networks—we say simply that spacetime is realized by a plurality, which is nothing more than a group of basic components, not cashed out in mereological terms.

The response still faces two problems, however. The first problem is to say what pluralities are. We know what they are not: they are not mereological wholes. But we still need some sense of what they are, so that we can fully understand the picture of spacetime emergence at issue. What we need is a non-mereological notion of a whole. While it is not completely clear how such an account might go, let us grant for the sake of argument that we know what pluralities are, or that some such account can be given.

The second problem is to specify *which* pluralities play the relevant functional role. Not just *any* plurality of neurons can realise a mental state, only very specific pluralities can do this work. For instance, the plurality containing one neuron from every person on earth can't plausibly play the functional role of any mental state. Rather, what is required is a plurality of neurons that are wired in a certain way. But what way is that? Presumably, it is only pluralities of neurons that come together to form a single, integrated network that plays the role of a mental state. But then it is very tempting to think we've just re-invoked neural states in order to individuate the pluralities that play the right functional role. It is only pluralities of neurons that form a single, integrated neural structure that can realise mental states. The question then arises, once again, of what the metaphysical relationship might be between individual neurons and larger neural structures consisting of pluralities of neurons that enables the larger structure to play a certain functional role. Here it is, once again, very tempting to call on mereological or constitutive language to cash out the relationship at issue.

So too, in the case of spacetime. One can certainly call on a plurality of basic components and say that the plurality realizes spacetime. But only some pluralities can do this work. For instance, a random plurality of nodes and edges from a spin-network, for example, simply cannot realize spacetime. But then the question arises as to what individuates those pluralities of nodes and edges in a spin-network that can realise spacetime from those that cannot. Which pluralities can realize the spacetime role? The answer, it would seem, is those pluralities that form weave states. But now we need an account of what the metaphysical relationship might be between the components of a weave state and the weave state itself. As before, it seems very tempting to reintroduce mereological or constitutive language. And the difference, again, with the neural case is that this seems fine when we are talking about mental states, but less than fine when we are talking about spacetime.

Of course, there could be some structuring principle that allows one to explain the relationship between weave states and components of a spin-network that is non-mereological or non-constitutional in nature. We cannot rule that out, though we don't see what such a principle might be, though perhaps it makes sense to wait for developments in physics to see if some hint can be gathered there (more on this at the end of the chapter). However, if such a principle can be found, then as above it does tend to weaken the need for spacetime functionalism. We could just use this structuring principle directly to explain the metaphysical emergence of spacetime.

6.7. Another Entity?

Before wrapping up, it is important to consider a broad objection to the arguments presented in this chapter. We have focused exclusively on the metaphysical emergence of GR spacetime. One might worry, however, that in doing so we have made matters easier for ourselves. The emergence of that specific entity is, in general, quite difficult to accommodate. If, however, we focus on the emergence of some *other* entity, then perhaps our arguments can be avoided. For instance, one might argue that *time*—in the form of a C-series—emerges directly from the fundamental structure posited by a theory of QG. Alternatively, one might argue that some entity *other than* GR space-time emerges, and this entity is sufficient for the existence of a C-series.

We don't think that this response will succeed. The arguments we have presented in this chapter are likely to generalise. The accommodation of any emergent temporal structure will require providing an account of the emergence of time from something that is not temporal (assuming, of course, that theories of QG lack spatial, temporal and spatiotemporal structure fundamentally). The difficulties we have raised for the metaphysical emergence of spacetime apply equally well to the emergence of time. In whatever way time emerges, there must be some metaphysical relation in virtue of which it emerges. The options available to us are composition, constitution and functional realisation. The first two relations are conceptually linked to space and time in a manner that prevents them from being used to account for the emergence of time itself. The third relation would either force temporal structure back down into the fundamental ontology of a theory of QG or push us back to one of the first two relations. We thus don't see the prospects for the emergence of a C-series as being any better if we shift our focus away from the specific ontology of GR.

6.8. Conclusion

In this chapter we have considered, in detail, the metaphysical emergence of GR spacetime. The metaphysical emergence of spacetime looked as if it could be one way to resist the quick argument for timelessness outlined in

Chapter Five. If spacetime must metaphysically emerge, then it must exist, and, given that it exists, so too does a C-series. We have argued, however, that there is no viable account of the metaphysical emergence of GR spacetime available. The very notions that we might use to elucidate the emergence of spacetime, or for that matter time, from some non-spatial, non-temporal, non-spatiotemporal structure, only really make sense within a spatial, temporal, or spatiotemporal context, and stretch to breaking point when applied to space-time or space and time.

It is important to be clear about what we take ourselves to have shown. Recall the dialectic: the quick argument for timelessness aims to establish the epistemic possibility that the C-series does not exist. The metaphysical argument from emergence is supposed to be a way of closing this possibility. For the argument to do this, however, it really needs to show that spacetime *must* emerge, epistemically speaking. We take ourselves to have defeated this modally strong assumption. Spacetime may not be emergent, since there may not be any viable metaphysics of this kind of emergence. We have argued for this claim on the grounds that current tools within metaphysics for understanding spacetime emergence appear to break down, and in quite a general fashion.

It is, of course, compatible with everything we've said here that spacetime *could* be an emergent phenomenon. There may well be a way to develop a viable metaphysics of this kind of emergence. One option might be to appeal not to the specific relations we have discussed here, but to analogous relations that are fit for purpose. So, for instance, one might accept that mereological relations are no good, but maintain that there are relations that work better that are analogous to mereological relations in important ways (*mutatis mutandis* for the other options considered here). One could also combine this with a grounding-based account and use these analogical parthood relations as a basis for explicating how grounding works in the case of spacetime emergence. For instance, it might be that such relations are ultimately responsible for grounding various aspects or parts of an emergent spacetime structure.

We are amenable to this line of thought. Indeed, given the kind of radical shift in how we think about physics on the horizon in QG, one might well expect that there should be a corresponding reimagining of our standard metaphysical tools for structuring the world. Thus, one might not find it all that surprising that mereology of the ordinary kind fails to apply in this new, and wild regime. We thus think it is very much worth considering whether spacetime emergence can be made to work. All we hope to have shown is that the success of the emergence program is by no means guaranteed. For by the very same token, the emergence of spacetime might well fail because dependent spacetime cannot be cashed out in metaphysical terms. As matters stand, we can't really say confidently either way: that spacetime will or won't be emergent. But this uncertainty is all we need to save the quick argument for timelessness from the challenge posed by metaphysical emergence. Since it is only the epistemic possibility of a world without time that is needed, and our inability to say that spacetime will be shown to metaphysically emerge is enough to secure that possibility.

In the next chapter, we consider a response to the relatively weak conclusion that we take ourselves to have established. Even if we cannot provide a viable metaphysics of it just yet, nothing particularly deep follows. All we have shown is that our current metaphysical resources are not up to the task. But that is just a limitation of philosophy as it stands and does not provide a reason to doubt the existence of spacetime. Thus, we should continue to believe that spacetime is metaphysically emergent *even if* no good metaphysics can be given. Our aim is to provide a response to this style of argument and to thus complete our case in favour of the quick argument for timelessness.

7

Approximating Spacetime

7.1. The Problem of Empirical Coherence

In the previous chapter, we argued that spacetime might not be metaphysically emergent in QG on the grounds that current approaches to emergence seem to break down for spacetime, and so spacetime emergence may end up being a failed idea. One may not be particularly swayed by this argument. There is, according to several philosophers, serious pressure to accept the existence of spacetime. We thus have very good reason to suppose that spacetime must be metaphysically emergent *even if we can't see how.*

In this chapter, we present the central argument toward this conclusion. The argument is based on a cluster of three problems for theories of QG, problems that are based on a perceived problem to do with empirical confirmation for such theories. Having set out three related versions of the problem, our aim in this chapter is to then solve two of them. In the next chapter, we'll deal with the third. The overarching goal here is to find a way to make sense of empirical confirmation in QG without positing an emergent spacetime. If we can do that, then we have no reason to feel the force of the pressure described above, to suppose that spacetime *must emerge* even if we don't know how.

7.2. Three Problems of Empirical Coherence

Theories that deny the fundamental existence of GR spacetime and spatio-temporally located entities face a threat from empirical incoherence.[1] A theory is empirically incoherent when the truth of that theory undermines our

---

[1] It is important to note that we mean something very specific by 'spacetime'. We have in mind the particular entity that is posited by GR. We concede that there may be a looser sense of the word 'spacetime' that is used in philosophy. This looser sense of spacetime is more or less synonymous with the concept of physical location. As we shall see later on, there is reason to pull 'location' and 'spacetime' apart, since it may be that there is a viable notion of location that is not spatiotemporal in nature. Thus, for present purposes we will focus on a more technical conception of spacetime.

Out of Time: A Philosophical Study of Timelessness. Sam Baron, Kristie Miller, and Jonathan Tallant, Oxford University Press.
© Sam Baron, Kristie Miller, and Jonathan Tallant 2022. DOI: 10.1093/oso/9780192864888.003.0007

empirical justification for believing it (Barrett 1999). As we see it, the threat from empirical incoherence can be decomposed into three distinct, but related problems.

First, the *problem of location*. The problem is this: observations are observations of (a suitably generalised understanding of) what are sometimes calls 'local beables'—very roughly, 'things that are situated in spacetime'. Thus, if spacetime and spatiotemporally located entities do not exist as a part of a fundamental theory of QG, then such a theory seems to imply that there are no local beables and, as a consequence, that there can be no observations that support the theory.[2] Any observational support for the theory would thus imply that spacetime and spatiotemporally located entities exist, which would undermine the theory itself.[3] In order to avoid the threat from empirical incoherence, then, we must assume that spacetime exists—or so the thought goes.

Second, there is the *observation problem*, outlined by Healey (2002). This problem focuses on *causation*, rather than *location*. Causation, it seems, is necessary for empirical confirmation to occur. In order to be able to observe a local beable, it is not enough that it be located. It must also be the case that an observer is capable of causally interacting with the local beable in question. It is commonly thought, however, that time is necessary for causation. As noted, theories of QG are supposed to be ones in which space, time and spacetime are all missing. It would seem, then, that such theories cannot support causation, and so cannot support observation.

The third and last problem of empirical coherence is what we will call *the problem of content* (Braddon-Mitchell and Miller 2019). In order to be able to accept a theory of QG as true, one must first be able to represent the theory to one's self. Mental representation, however, presupposes the existence of mental content. On standard naturalistic theories of mental content, there being any mental content at all relies on there being causal connections between mind and world. As before, if causation requires time, then it would seem that theories of QG that eliminate time also eliminate causation and, with it, mental content. But then, fairly plausibly, timeless theories are empirically incoherent in the following sense: one cannot come to believe any such theory (even on non-empirical grounds) without presupposing the very structure that the

---

[2] We focus on the problem as it arises in QG. However, a similar difficulty arises for configuration space realist interpretations of quantum mechanics. See Maudlin (2007) and Ney (2015) for discussion.
[3] This statement of the problem is based on Huggett and Wüthrich (2013).

theory in question rejects; the very act of believing the theory involves representing it, and if there is such a representation, then the theory must be false.

Although both the problem of content and the problem of observation raise essentially the same issue—the need for causation—we take them to be separate concerns. The problem of content arises earlier in the confirmation process. One must represent a theory to one's self before one is able to apply any information gathered from empirical confirmation to produce a judgement concerning the theory in question. As we shall see, it is also possible to solve the problem of content without thereby solving the problem of observation, and so it makes sense to pull them apart in the first instance.

In what remains of this chapter we will proceed in three stages. We will begin by dealing with the problem of location. To do this, we propose a spacetime anti-realist interpretation of QG. After that, we will turn to the problem of content, and use the same broad anti-realist approach to recover content as well. We conclude by considering an objection against the anti-realist position developed here. We defer the problem of observation to the next chapter, where we take a closer look at time and causation. Collectively, these discussions will enable us to reject arguments toward the necessary emergence of spacetime.

7.3. Location, Location, Location

The problem of location relies on a particular understanding of what it is to be a local beable, along with a particular understanding of the nature of empirical confirmation. Specifically, the problem presumes that entities must be spatially, temporally, or spatiotemporally located in order to be observable. Indeed, local beables are partly defined in terms of their spatiotemporal locations.

Our solution to the problem of location is thus straightforward: theories of QG should be interpreted as requiring a new non-spatial, non-temporal, non-spatiotemporal conception of location. In short, the loss of space, time, and spacetime, should not be interpreted as the loss of location.

This notion of location is not obviously incoherent. Philosophers have long taken seriously the prospect of entities being located, but not in space or time. For instance, mathematical Platonists maintain that mathematical objects exist, and are located 'outside' of space and time, in some shadowy abstract realm. Similarly, many philosophers believe that God exists, but unconditioned by space and time, located in God's own realm that stands apart from both. Heaven itself is thought to exist beyond space and time in some sense; so too for Nirvana. Lewis's modal realism is also a case in point: concrete possible worlds are located in some broader modal space that is not itself spatial or temporal in nature. The similarity measure that Lewis throws over his worlds also inducts them into another kind of space, often used in philosophy: location in a similarity space.

Other examples abound. For instance, one might speak of the colour red being located in a colour space; or of a theory occupying a position in logical space. In both cases, we make these claims because there is some imagined abstract space, which allows us to individuate distinct locations within it, even though those locations are not spatial or temporal in nature. In physics, it is commonplace to speak of location in a configuration space or a phase space, neither of which involve anything like space or time, but come complete with a great deal of structure. Typically, what we find in these spaces is a metric of some kind, with which we can speak meaningfully of the distances between points. The metric function is not a spatial or temporal function.

It is tempting to treat these notions of location as metaphorical, but we don't see why one should be forced to do so. Instead, we take these examples at face value, as involving the literal use of a concept of location. We take them as evidence that location is a broad notion, one that is structural at its core. So long as there is enough structure to be able to produce a distance function, then we have location. Since there are very many distance functions that are not spatial or temporal in nature, there is no reason to suppose that location is essentially tied to these notions.

Now, one could hold the view that entities in QG are located in the same 'realm' as mathematical objects (or any of the other notions just discussed).[4] This, however, is not our view. Indeed, we take it that whatever location in QG might be, it is not one of the ones just articulated; it is something new. What is it? Well, to a certain extent, that is a matter for physics. Whatever the fundamental physical structure might be that is posited by our best physical theories, location is *location within that structure*. If the structure is not spatial, temporal, or spatiotemporal, then neither is location. So, for instance, in the case of LQG, we should just take the fundamental physical structure defined by a spin-network to define a notion of location.[5] Location, on this view, is just

---

[4] The Pythagorean view of the universe as one that is made of mathematical objects, and thus essentially mathematical in nature, is a historical precedent for a view along these lines.

[5] It is important to keep in mind, however, that the locations of entities within a spin-network don't line up with their apparent spatiotemporal locations. This is because of the problem of *disordered locality* (cf. Huggett and Wuthrich 2013). Entities that are adjacent in a spin-network—i.e., that are located at vertices that are connected by some edge—generally correspond to regions that are arbitrarily

location within the physical structure represented by a spin-network. For other theories, such as the theory advocated by Barbour (1999; 1994a; 1994b), the fundamental structure is a configuration space, and so location on this view is location within whatever physical structure the configuration space represents.

What we are recommending, then, is what Huggett and Wüthrich call a 'top down' approach to location. In a 'top down' approach to location we assume that a physical theory supplies locations for entities so long as it meets certain minimal requirements—namely, providing enough structure to be able to produce a distance function over entities. We then let the theory tell us how location works and judge the theory's confirmational status by its own lights, in terms of observations of entities that are located in the sense proposed by the theory. Thus, in the case of QG, we start from the assumption that a theory like LQG has enough structure to provide locations, but we don't take the theory to task for failing to provide the specific kinds of locations that we find in GR (or, indeed, any other theory).[6]

Now, one might worry that our 'top down' approach to location leaves the relevant notion of location dangerously unspecified. We have said that the notion of location operative in QG should not be interpreted as spatial, temporal, or spatiotemporal. However, we have also denied that it should be interpreted along the lines that philosophers have previously thought of similar notions of location, such as the kind that is appropriate for mathematical objects or for God. We have not said very much to elucidate this particular conception of location. This is a problem, one might argue, because we have offered no assurance that the notion of location at issue can in fact do the work required of it to support empirical observation and confirmation.

However, the notion of location is not completely unspecified. As indicated, we need enough structure for a distance function to be definable, and so the notion of location, we suggest, requires at least a connection to distance in this

---

far away from one another when considered under a spatiotemporal description. What this means is that the locations of entities at high energies does not, in general correspond to their location at low energies.

[6] A 'top down' approach is to be contrasted with a 'bottom up' approach. In a 'bottom up' approach to location, one comes to a physical theory with an existing concept of location, which is then used as a standard by which to judge the confirmational status of the theory in question. In the case of QG, this involves coming to a theory of QG with a spatiotemporal conception of location, and then using that as a basis upon which to judge the theory. We think this is a mistake. It is analogous, we maintain, to judging a spatiotemporal theory using a classical Newtonian conception of location. Imagine, for instance, complaining that GR is empirically incoherent because it does not feature space and time as they appear in a Newtonian theory. Such a response fails to take GR seriously. So too, in the case of QG, we maintain that holding a theory of QG to the locative standards of GR fails to take the former theory sufficiently seriously.

minimal sense. Further, there is a bit more that we can say, drawing on the discussion of functionalism and theoretical emergence considered in the previous chapters. There is a limitation to the style of functionalism advocated by Lam and Wüthrich. Recall that, according to this brand of functionalism, a functional analysis of the specific entity posited by GR is provided by producing a Ramsey sentence for that theory. This will, at a minimum, involve providing a functional analysis of the metric field in GR. The trouble is that having produced a functional analysis along these lines, it is unclear whether there is anything that can satisfy the Ramsey sentence. One dimension of the problem relates to discreteness. The metric field in GR is continuous, and yet there is a widely held expectation that the fundamental structure of QG will be discrete.

At best, the relationship is one of approximation. Indeed, this is often how physicists view the relationship between QG and the metric field of GR: nothing in QG exactly corresponds to the metric field of GR, but there may well be something that mimics it, under certain conditions. Specifically, at low energies, the structure posited by a given theory of QG approximately realises the metric field of GR. What this means, we take it, is that the behaviour of the physical structure posited by a theory of QG matches the physical behaviour of the structure posited by GR sufficiently closely within a certain regime, that the two structures make effectively the same empirical predictions. Thus, despite the fact that the fundamental structure in QG is discrete, we can safely ignore the differences between a discrete structure and a continuous metric field for the purposes of doing physics, so long as the energy level in whatever system we are attempting to model is sufficiently low.

Our discussion of theoretical emergence gives us good reason to suppose that any theory of QG ought to be able to reproduce the empirical results of GR in much the same way. It is thus a general constraint on a theory of QG, then, that it at least approximates the metric field of GR at low energies, regardless of what the specific physical structure being posited at the fundamental level might be.

This gives us a great deal of information about the notion of location being posited by any given theory of QG. We know that whatever location amounts to within such a theory, it must *at the very least* approximate location as it appears within GR spacetime, by approximating the metric field. Moreover, the approximation should be sufficiently close that whatever empirical tests one might dream up for GR will hold good under the theory of QG as well. Thus, we are not entirely in the dark about our non-spatial, non-temporal, non-spatiotemporal notion of location. We know at least this much: at low energies, location in this sense is empirically indistinguishable from locations determined by the metric field of GR.

Of course, this doesn't tell us what location is like at high energies, and it doesn't shed much light on what location is like essentially, if it's not spatial, temporal, or spatiotemporal. But it does tell us enough to know that the problem of location can be resolved. What we know is that location in QG behaves in much the same manner as location in GR, when it comes to empirical prediction, observation and confirmation. Thus, so long as location within GR can provide a basis for empirical confirmation, the same should apply equally to QG.

Now, one might worry that the existence of something that approximates spacetime poses a problem for any claims to the effect that time does not exist (if we approximate spacetime too well, might we not in fact reintroduce time?) We will return to this issue below. First, however, we will turn to the second problem of empirical coherence: the problem of content.

7.4. The Problem of Content

As discussed, mental content seems to presuppose a causal connection between mind and the world. Indeed, most naturalistic theories of content fall into four broad categories: causal (Dretske 1981; Fodor 1987; 1990; Stampe 1986); sophisticated co-variational (Maloney 1994; Dretske 1983); teleonomic (Millikan 1984; 1989a; 1989b; Neander 1991; 1995; 1996); and functional role theories (Field, 1977; Block 1986; Horwich 2005). On all of these views it is necessary (though not sufficient) that certain kinds of causal relations obtain between a mental state and the represented state. Very roughly, according to causal theories, what a state represents is what causes that state. According to functional role theories what a state represents is a function of both what causes that state, and what that state causes. According to teleonomic theories, what a state represents is a function of the evolutionary history of that state. Sophisticated co-variational accounts require there to be a reliable connection between the mental state and what it represents; this might be a causal connection, or it might involve a complicated story about counterfactual dependence between the two states. For now, let's set aside talk of counterfactual dependence and focus just on those accounts that appeal to causal relations.

If causation requires time, and if there is no time, then there is no causation either. Thus, any account of content that appeals to causation will be one according to which a timeless world is a world that lacks mental content.

On the face of it, the problem of mental content imposes a requirement to recover causation from an underlying ontology that eliminates space, time, and spacetime. One response to this worry, then, is to show that QG worlds do not lack causation, though they do lack time. Another response is to show that an account of mental content can be given that does not presuppose causation. In what follows, we offer an account of the second kind. There are, in fact, a range of such accounts already available. We canvas these and find them wanting, before offering our own.[7]

7.4.1. Four Proposals

Naturalistic theories of mental content appeal, *inter alia*, to reliable connections between the mental state doing the representing, and the state of the world being represented. If we cannot appeal to causation between one state on the other, then another option is to appeal to structural isomorphisms between brains (we assume here that mental content resides in brains) and states of the world. Why does a particular brain state represent some particular worldly state? It does so, according to this proposal, because the brain state is structurally isomorphic with that state of the world.

The problem with this approach is that there seems little reason to suppose that our brain states *are* structurally isomorphic with states of the world. There are no structural isomorphisms between the derived non-mental contents of books, or paper records, or indeed, computer records, and the parts of the world that we suppose those records to represent. There are no structural isomorphisms between temporal durations and tree rings. Indeed, we can expect that records will often fail to share any structural isomorphisms with the things about which they are records because efficient ways of storing information (as in brains, or computers, or books) are typically ways that do not preserve isomorphisms. So, while this proposal offers the prospect of vindicating mental content without causation, it does not seem to us to be a very plausible way to do so.

A second option, then, is to appeal to synchronic relations that obtain between the represented state and the representing state. There are already accounts of mental content that seem to go at least some of the way towards offering such an account. Inferential (or conceptual) role semantics,[8] for

---

[7] For further discussion of these issues see Braddon-Mitchell and Miller (2019).
[8] See Harman (1982); for a more sceptical take see Lepore (1994).

instance, holds that the content of mental states is given by their inferential roles. Thus, one might think that what it is for a mental state to have the content *conjunction* just is for that mental state to license certain inferences.

Like any philosophical theory of content, inferential role theories have their problems. In what follows, however, we won't focus on those worries. Instead, we are interested in whether it is possible to account for mental content without appealing to causal relations by instead appealing to such synchronic relations.

The answer is that this is not clearly possible. Typically, inferential role theorists hold that inferential roles are at least partially determined by those inferences a subject makes. The inferences a subject *makes*, however, involve causal processes; inferences are constituted by a causally connected sequence of mental states. That being so, traditional versions of inferential role theory are of no help, since they require the existence of causation.

Still, that opens the possibility of developing an entirely synchronic version of inferential role semantics. To do this, one would need to say that inferential roles are exhausted by instantaneous dispositions, rather than requiring any actual diachronic inferences. In turn, we would need to make sense of subjects having instantaneous dispositions to make certain inferences, even though there are no such inferences, conceived as causal processes. It remains unclear, however, whether we can adequately make sense of inferential roles by appealing to instantaneous dispositions (certainly extant inferential role theories do not take this route) or whether we can make sense of instantaneous dispositions in the absence of both spacetime and causation. So, while we cannot entirely rule out such an account, we remain sceptical of this approach.

The third proposal is to suppose that representational content is grounded in phenomenal consciousness, and that the latter can obtain in the absence of causal connections between mind and world. To get clear on such a view, let's call the way an experience feels to the subject, that is, the *what it is like* to have that experience, its *phenomenal character*. Then, feeling pain, or seeing redness, or hearing something loud, are all examples of phenomenally conscious experiences: conscious experiences with phenomenal character. According to reductive representationalist views, phenomenal character reduces to representational content. According to such views, having an experience with a particular phenomenal character is grounded in having an experience with a certain representational content (Harman 1990; Tye 2000; Chalmers 2004b).[9]

---

[9] On views like this, if there is no representational content, then there are no phenomenally conscious experiences. So, views like this make additional trouble for timeless theories. If it turns out that we cannot make sense of representational content, then it also turns out that we cannot make sense of phenomenally conscious experiences.

For our purposes, what matters is that we can invert the order of explanation here. According to phenomenal intentionality theorists, there is something called *phenomenal intentionality,* which is intentionality that is constituted by phenomenal consciousness. (Here, 'intentional states' is just another name for representational states). So, a phenomenal intentional state is a representational state that is constituted by phenomenal states (Pautz 2013; Kriegel 2011; 2013). That doesn't tell us anything about the *direction* of explanation and is consistent with reductive representationalism about phenomenal consciousness. But we can combine this idea with the further thought that representational content reduces to phenomenal consciousness (rather than the other way around). Then, if one holds that it is possible to have phenomenally conscious experiences in the absence of causal connections between those experiences and the world, one has an account of mental content that does not require that causal relations exist. Moreover, this last claim is often taken to be plausible.

To illustrate, consider a brain in a vat. We can imagine that the brain is a phenomenal duplicate of someone, Brian. Things seem the very same to the brain in the vat, as they do to Brian. The brain in the vat is not causally connected to any of the things in the world to which Brian is connected. Hence many of the brain's beliefs are false. But, one might think, not only do things seem the same to the brain as to Brian, but, in addition, the brain represents things to be the very same way as Brian does: it is just that the brain is wrong about how things are. But if that is right, then it cannot be that representational content is a matter of bearing some connection to the world, since the brain has mental states with the same contents as Brian does. Plausibly, then, it must be a matter of having phenomenally conscious states. It is because Brian and the brain are phenomenal duplicates that they are also representational duplicates (see Loar (2003) and Horgan, Tienson, & Graham (2004)).

There are, however, some important things to notice about the view of phenomenal intentionality that would be required to do the work necessary here. If such an approach is to capture the wide variety of representational states that we take ourselves to have, then we will need to sign up for a very strong version of the phenomenal intentionality theory according to which either all representational states can be reduced to phenomenally intentional states, or, at the very least, the content of all representational states is fully grounded in phenomenally intentional states. The view that all representational states can be reduced to phenomenally intentional states is controversial (though see Pitt (2004), Farkas (2008), and Mendelovici (2010) for a defence of the view). It is relatively easy to see how it could be that a representation of some object in the world—a blue cube for instance—reduces to there being a phenomenally conscious experiences of a certain shape and colour. It is much less easy to see how complex thoughts about the world—that, for instance, a particular theory of QG is true—could consist in there being some phenomenally conscious state.

Instead, one might concede that not all representational states reduce to phenomenally conscious states. Non-phenomenal intentional states, for instance, do not. Nevertheless, if the content of every non-phenomenal intentional state is in some way fully grounded in phenomenally conscious states, then this will do just as well in vindicating mental content. How could the content of non-phenomenal intentional states be grounded in phenomenally conscious states? One suggestion is that non-phenomenal intentional states are dispositions to have phenomenal intentional states. These dispositions get their contents from the phenomenal intentional states that they are dispositions to bring about (Searle 1983, 1992, 1997). So, a belief that P, which has non-phenomenal intentional content C, gets that content by being a state which is disposed to bring about a phenomenal intentional state with content C. Hence, although there are contentful states that are not reducible to phenomenally conscious states, nevertheless, all the *content* of states is ultimately reducible to phenomenal consciousness.

Even this view, though, seems like a bit of a reach. On this view there must be some non-phenomenal conscious state which has a particular content, namely the content as of some particular theory of QG being true. That state gets its content from some phenomenally conscious state that it is disposed to bring about. However, it is unclear what the phenomenally conscious mental state at issue might be. Quite generally, it's not clear that phenomenal character is sufficiently rich to ground the very fine-grained contents of representation that we in fact seem to have. Still, so far this seems to us to be perhaps the best option on the table. We certainly don't want to suggest that we have shown that such an approach must fail. Nonetheless, for the reasons given, we think it can be tentatively rejected.[10]

That leaves us with our fourth option, which is to consider some kind of non-naturalistic approach to mental content. These theories do not appeal to causal relations between mind and world. According to non-natural theories of representation, mental states do not come to have the content they do

---

[10] Note that by claiming that phenomenal character is insufficiently rich, we are not claiming that there can't be cognitive phenomenology as of a particular theory being true. Our claim is compatible with there being phenomenology connected to cognition (rather than perceptual experience). The point is that *even if* we admit cognitive phenomenology, we'd still need a rich array of phenomenal states and it's not clear that phenomenal life is sufficiently rich.

because of some naturalistic connection they bear to the things they represent. Representational Platonism is such a view.[11] On that view, mental states have the content they do in virtue of bearing some relation to the Platonic forms; these forms are, as it were, the content, and mental states get their content by *sharing* in the forms. Thus, one's mental states could have the very same content they do, even if the world were very different from what it is.

Another version of non-naturalism is a kind of primitivism about content. On such a view, it is simply a brute matter that our mental states have the content they do. Any of these sorts of non-naturalism will do the job here. Indeed, since these theories don't really require any connections between mind and world, there is really nothing a theory of QG could entail, that would undermine such a theory of mental content. The problem, of course, is that non-naturalistic theories are unattractive. Even if it's not entirely clear, in virtue of what Annie is currently representing, that there is a ball, it seems very odd to think that it's because her mental state shares in some form of a ball. So, while this option is certainly one that the QG theorist could adopt, it can hardly sit well if this is the best option going.

So far, we have explored and tentatively rejected four options that attempt to account for mental content in the absence of any causal connection obtaining between minds and worlds. At the beginning of this section, we noted that while most naturalistic accounts of mental content appeal to causal connections, some appeal to relations of counterfactual dependence between mental states and the world. Indeed, some co-variational accounts are explicitly framed in terms of counterfactual dependence. Now, it seems clear that one can have relations of counterfactual dependence without having relations of causation: even those like Lewis (1973) who analyse causation in terms of counterfactual dependence will allow that there can be counterfactual dependence without causation (though not, of course, causation without counterfactual dependence). So, an obvious next port of call is to consider the prospects of developing a naturalistic account of mental content that appeals to relations of counterfactual dependence.

7.4.2. Mental Content and Counterfactual Dependence

We've just said that, at least in principle, one can have counterfactual dependence without causation. On its own, counterfactual dependence is not enough

---

[11] For a useful survey of such views see Cummins (1989).

to produce an account of mental content without causation. For we must ensure that the *right* counterfactual dependencies are preserved. It needs to be that our mental states counterfactually depend on the right states of the world. Moreover, it must be shown that this can be so in the absence of space, time, and spacetime. Can this be done?

We believe that the answer is 'yes' but admit that the issue is a difficult one. Here we hark back to our solution to the problem of location. Suppose for a moment that GR spacetime exists. The existence of GR spacetime, we assume, helps to support certain relations of counterfactual dependence. If one believes in the existence of spacetime, then one has a route to counterfactual dependence. If, however, spacetime does not exist, but something *very similar to it does* and, indeed, something that behaves in the same way as spacetime does at low energies, then we have some assurance that the same or at least very similar counterfactual dependencies will hold for this non-spatiotemporal structure as well. To put the point another way, the differences between existing spacetime and a structure that approximates spacetime to a high degree but that behaves in the same manner as spacetime in a certain regime are not sufficiently great to undermine counterfactual dependence.

Of course, it may not be the case that the same counterfactual dependencies are upheld between the two pictures: one in which GR spacetime exists, and one in which something that approximates GR spacetime exists. Certainly, at *high* energies, when the behaviour of the QG ontology comes apart from the behaviour of spacetime, we can expect very different counterfactuals to be true (and maybe even the loss of counterfactual dependence more generally). But with respect to the low energy regime in which mental states, observation and empirical confirmation standardly occur, the counterfactual dependences are, if not the same, at the very least extremely similar.

Here is another, more abstract, way to put the point. Counterfactual dependence, in its barest form, requires two things. First, it requires a structure of some kind, in which there are connections between aspects of the structure. Second, it requires rules over the structure that dictate how one part of the structure changes, when an alteration is made to some other part. For counterfactual dependence to fail *outright*, one or both of these features must be missing. Now, while we may lose the existence of spatiotemporal structure in QG, and the full generality of the laws of GR, we do not lose structure *tout court*, nor do we give up on the presence of structure-governing laws of some kind. Rather, what we find is just that, *fundamentally*, the structure is not a spatiotemporal one (though it is still a structure of some kind, such as a spin-network) and the laws that govern the structure are not the laws of GR; they are the laws of the relevant theory of QG. This is, essentially, to reiterate the point that we made about location: it is precisely because there is a great deal of structure remaining in theories of QG that we have the basis for a notion of location. So too, we submit, do we have the basis for counterfactual dependence, so long as the structure is governed by laws that inform about what happens to one region of the structure, given an alteration to another region.

Assuming the structured nature of QG, there is good reason to suppose that at least some counterfactual dependencies remain. Although the laws of QG are different from the laws of GR, GR is supposed to be derivable from the QG laws, under certain conditions: namely, at low energies. It follows, then, that there is a regime in which the QG laws operate in the same manner as the GR laws—specifically, low energy regimes. Moreover, under the same conditions, the ontology of the underlying theory—the structure that supports counterfactual dependencies—is sufficiently close to spacetime that it operates in accordance with the GR laws. Given the close parallel between the structure and the laws of spacetime and the structure and the laws of the QG ontology within the relevant regime, very similar counterfactuals will be true of both structures within the limited domain where the structures accord with one another.

Of course, there is room for some variation: the QG ontology only approximates spacetime. But the differences are small enough to be safely ignored from the perspective of physics within the relevant regime, and so, we submit, the differences are sufficiently small that they can be ignored when it comes to the counterfactual dependence of minds on the world as well. If that's right, then because mental states and mental content all exist in the regime where QG and GR agree, it is reasonable to suppose that the counterfactuals needed to support mental representation are available in both cases.

Thus, we think that the best way forward is to argue that the presence of these dependencies, even in the absence of causation, is sufficient for mental content. What naturalistic theories need is that the presence of some mental state is a reliable indicator of some state of the world. A very natural way for this to be, is for that mental state to (typically) be caused by that state of the world. But reliable indication need not involve causation. If one state counterfactually depends on another, then the former is a reliable indicator of the latter. That, we maintain, is enough for mental content.[12]

---

[12] The idea that there can be reliable indication without causation is similar to a Malebranch-style view discussed by Ney (2012) in the context of wave-function realism in quantum mechanics.

7.5. Approximation and the C-series

So far in this chapter, we have provided solutions to two problems of confirmation that apply to theories of QG: the problem of location and the problem of content. In this section, we consider an objection to those solutions. To grasp the objection, we need to pull back a bit and recall the broader dialectic. We began, in Chapter Five, by offering a quick argument for timelessness. The argument sought to establish as an open epistemic possibility that the C-series does not exist, since it seems to be eliminated from a number of approaches to QG. We then raised a challenge to this argument: GR spacetime is an emergent phenomenon within QG, and so it exists. This, in turn, implies the existence of the C-series. We then argued against the claim that the theoretical emergence of spacetime implies its existence and, subsequently, argued against the claim that spacetime must be metaphysically emergent. In this chapter, we have been considering an argument that spacetime must be metaphysically emergent, despite everything we have said in Chapters Five and Six. Spacetime must be emergent because if it is not, then theories of QG will be empirically incoherent.

As we saw, the problem of empirical coherence is really a cluster of three related problems. In response to the first two problems—the problem of location and the problem of content—we have suggested that there is enough structure in a theory of QG to support empirical coherence, though it is not spatial, temporal, or spatiotemporal structure. While this structure may not be spatial, temporal, or spatiotemporal, it is sufficiently complex to support a range of different phenomena. It can support both location and content via counterfactual dependence. Moreover, we have confidence that it can support both in exactly the manner needed to support confirmation because the structure itself approximates GR spacetime sufficiently well under the conditions in which empirical observation occurs.

The worry, then, is that our solution to the first two problems of empirical coherence may ultimately undermine the quick argument for timelessness. In this section, we consider and reject five reasons to think that this is correct. We also look at a nearby problem, which relates back to the empirical work described in Part One.

First, if something that approximates spacetime sufficiently closely exists then, while it may not follow that GR spacetime exists, perhaps such an entity is nonetheless sufficient for the existence of a C-series. After all, if the structure at issue provides locations for entities, locations that ground empirical observation, then one might think that it just follows that there exists at least a C-series. That would be too quick, however. As we have argued, location does not entail anything temporal or, for that matter, spatial, and so this first reason is unconvincing.

Second, one might retort that the presence of metric structure, which we have noted is likely to be found within the fundamental ontology of QG in some form (and that forms the basis of distance and thus location), is sufficient for the existence of the C-series. But, as before, there is no direct implication from the presence of metric structure to the existence of anything temporal. Plenty of metrics are not temporal in nature and so there is little reason to suppose that metric structure alone is sufficient for the existence of the C-series.

A third consideration relates to counterfactual dependence. We have argued that there is enough structure in a theory of QG to ground counterfactual dependence and, indeed, to ground more or less the same counterfactuals that are grounded by the existence of GR spacetime. One might argue that counterfactual dependence is sufficient for the existence of the C-series. But there is no reason that we can see to suppose that the presence of counterfactual dependence entails the existence of a C-series. So, there is no reason to suppose that, even if the presence of a C-series is sufficient for time, that the presence of counterfactual dependence entails the presence of temporal structure.

A fourth reason focuses on the upshot of the approximation. When something approximates GR spacetime it follows that a world with GR spacetime is empirically indistinguishable from a world in which some non-spatiotemporal structure exists that approximates GR spacetime at low energies. One might maintain that in any world that is empirically indistinguishable from a world with GR spacetime in some regime, a C-series exists. In short, empirical indistinguishability entails a lack of metaphysical differentiation, at least within the relevant energy regime where indistinguishability holds. This idea, however, seems to presuppose an unattractive form of positivism about metaphysics, according to which there can be no metaphysical difference without an empirical one. We reject this form of positivism, and so we reject the idea that the approximation at issue is sufficient for the existence of a C-series.

A fifth reason for taking the approximation to be sufficient for a C-series appeals to some brand of functionalism. At the low energy regime, the entity that approximates GR spacetime behaves in the manner required to support empirical confirmation. One might argue, however, that the C-series just is whatever is capable of supporting empirical confirmation in this manner. On this functional understanding of what a C-series is, it follows that the C-series exists wherever empirical confirmation is viable.

If the C-series is functionally analysed in this manner, however, and the C-series is supposed to be necessary for the existence of time in the folk sense, then it would seem to follow that some functional conception of time is necessary for time in the folk sense. But, as discussed in Chapter Two, we have evidence against the folk concept of time being a functional one. A functionalist conception is neither necessary nor sufficient for time in the folk sense. Indeed, the very functionalist conception we tested—one that focuses on our experiences of the world—seems like precisely the kind of concept one might need in order to argue from the existence of something that supports empirical confirmation to the existence of a C-series.

Perhaps there are further reasons to suppose that the existence of something that approximates GR spacetime implies the existence of a C-series, but we don't see what they might be. We recognise, however, that we haven't said a great deal about how the approximation works. This, one might argue, works against us. For clearly there are going to be some ways of approximating spacetime that are sufficient for the existence of the C-series. It really depends on the nature of the approximation.

Quite right. But this thought cuts both ways. Whether the existence of something approximating spacetime is sufficient for the existence of a C-series depends very much on the nature of the approximation at issue and, indeed, on the nature of the C-series. Granted, one could hold a view on which a C-series arises from an approximation of spacetime. But such an account, we contend, is not forced on us, at least not by the physics of QG. It thus remains open that the C-series does not exist, despite the fact that something approximates GR spacetime.

At this point, a further problem arises. Recall the purpose of the quick argument for timelessness. This argument is supposed to show that the discoveries that would lead people to deny that time exists are compatible with what we know about the world from science. It does this by showing that the loss of the C-series is epistemically possible, given our current understanding of QG, and so the loss of time in the folk sense is compatible with what we know. As discussed in Chapter Five, however, the loss of the C-series may not be what leads people to deny that time exists in the empirical work in Part One. There may be some other feature, F, missing from the vignettes that lacked a C-series that led people to judge that time does not exist. Thus, showing that the absence of the C-series is epistemically possible might not be enough to defeat the claim that the discoveries that would lead people to deny that time exists are incompatible with what we know from science.

We responded to this issue by noting that without an account of what F is, our opponent is not able to argue that the absence of F is incompatible with what we know from science in the first place. One might argue, however, that the appeal to approximate spacetime rather complicates this issue. The worry is that when presented with scenarios in which something approximates spacetime sufficiently well to reproduce the empirical results that support GR, people would in fact judge that there is time.[13] To semi-formalise

1. There is something, F (which is not identical to a C-series), the lack of which is triggering people's response that time does not exist in certain of the vignettes.
2. F is present in an approximation of spacetime.
   Therefore,
3. The folk will judge that time exists if actually there is an approximation of spacetime

This argument is sketchy. The first premise is nothing more than the empirical bet we discussed in an earlier chapter. We can't rule out there being some features other than a C-series that are salient to folk judgements about the reality of time. But we think we've some fair evidence that the C-series is what is doing the work and our opponent has nothing more to say here other than reiterating the possibility that something other than a lack of C-series is implicated in our judgements about the unreality of time in these vignettes. The second premise is unsupported. We don't yet know what F is. Our opponent cannot then insist that F is present in an approximation of spacetime. If there is something that approximates the C-series, then this may suffice for the existence of the very feature (or features) that lead people to deny that time exists (assuming that the relevant feature is not just the

---

[13] One intriguing possibility along these lines concerns the relationship between temporal ordering and metric structure. The C-series is a temporal ordering that gives rise to distance relations, and thus metric structure. One might argue, however, that the folk notion of time might be satisfied by something a bit weaker: a temporal ordering that does not have any metric structure. It might turn out that ordering alone is sufficient for the folk to agree that time exists. One might then argue that approximate spacetime gives us a temporal ordering without a metric. Thus, even though approximate spacetime doesn't yield a C-series, it nonetheless carries enough structure to satisfy the folk notion of time. Whether that's right, of course, depends on what the approximation looks like, and how it works. Accordingly, it is important to emphasise the point made above: namely, that we don't have a sense of the nature of approximate spacetime yet, and that's enough to leave it open that it won't preserve even temporal ordering.

C-series). However, it's equally true, in the absence of the details of the approximation, and in the absence of an account of what F is, that it may not. We really don't know, and so we don't detect a deep problem for our argument here.

There is a weaker version of the argument lurking, that is more persuasive but that doesn't impact our overarching conclusion. Simply, we can weaken both 1 and 2 to the claims that:

1. There could be something, *F* (which is not identical to a C-series), the lack of which is triggering people's response that time does not exist in certain of the vignettes.
2. F could be present in an approximation of spacetime.
   Therefore,
3. The folk could judge that time exists if actually there is an approximation of spacetime.

In this revised version of the argument, the conclusion has shifted to being that the folk *could* judge that time exists if actually there exists an approximate space-time. We can grant that. But since our overarching argument is simply that science doesn't rule out the discovery that time does not exist, the conclusion captured in 6 is no threat to us. That being so, it doesn't appear that there are good arguments against the position we've taken in this chapter.

7.6. Conclusion

We began this chapter with three problems of empirical confirmation for a theory of QG: the problem of location, the problem of observation and the problem of content. We have addressed the problem of location and the problem of content by appealing to a notion of location that is capable of underwriting counterfactual dependence sufficient for mental content. The problem of observation remains outstanding. We have left this problem to last because it really does seem to demand the existence of causation. It is difficult indeed to make sense of the existence of conscious observers without some notion of causation in hand.

In the next chapter, we will deal with the observation problem by arguing for two claims: that time is not necessary for causation and that a notion of causation without a C-series can be developed by leveraging the kinds of considerations outlined here, involving location, structure, and counterfactual dependence. In what follows we focus on providing an account of causation in the service of solving the observation problem. It's worth noting, however, that if one is unconvinced by our claim that mental content can be secured by appealing only to relations of counterfactual dependence, one can instead solve the mental content problem by appealing to the account of causation we present in what follows.

8

Causation and Time

8.1. Introduction

In this chapter, we consider the relationship between time and causation. Our goal is to offer an account of causation that does not require the existence of a C-series. The provision of such an account is motivated by the need to solve the observation problem: the problem of how we can observe entities in the world so as to provide empirical confirmation for a putatively timeless theory. Providing an account of causation along the right lines is challenging. One might argue that it is *prima facie* clear that without even a C-series there can be no causation.

In the first instance, then, we argue that a C-series is not necessary for causation. Our second goal is to sketch out an account of timeless causation. We show that the existence of something that approximates spacetime is sufficient to underwrite causation in the absence of a C-series. We begin, in Section 8.2, by outlining several arguments that might be used to establish a necessary connection between a C-series and causation, before responding to those arguments. Our aim, here, is to show that these arguments fail. We then offer a positive account of causation in Section 8.3. Our basic strategy is to develop a counterfactual theory of causation in the vein of Lewis's theory. While we focus on a counterfactual account, we will also gesture to how the basic strategy we endorse may be extended to other accounts of causation as well.

Before proceeding, it is important to offer a point of clarification. We accept a counterfactual theory of causation. Because, as we shall see, we can make sense of counterfactuals in the context of a timeless theory, we are tasked with defending the idea that there can be causation in a timeless world. Strictly speaking, however, counterfactuals are enough to do much of the work that we aim to do in this book. Counterfactual dependence appears to be sufficient to underwrite observation, and to thus address the observation problem outlined in the previous chapter. Counterfactual dependence is also enough for the picture of agency that we defend in Chapter Eleven. Those who defend a theory of causation according to which counterfactual dependence of the

Out of Time: A Philosophical Study of Timelessness. Sam Baron, Kristie Miller, and Jonathan Tallant, Oxford University Press.
© Sam Baron, Kristie Miller, and Jonathan Tallant 2022. DOI: 10.1093/oso/9780192864888.003.0008

kind we discuss here is not sufficient for causation, may have an easier road; they don't need to worry about the connection between causation and time quite so much. They need to worry only about (suitable) counterfactuals in a timeless world. For us, however, as believers in a counterfactual theory, matters are not quite so straightforward.

8.2. Causation and the C-Series

There are several arguments that might be marshalled to show that there is a necessary connection between the C-series and causation. These arguments can, however, be grouped into broadly two kinds: arguments from intuition (8.2.2—8.2.5) and arguments from theory (8.2.6—8.2.7). The first kind focuses on intuitions surrounding causation that seem to demonstrate the necessity of the C-series for causation. The second focuses on the way in which the necessitation claim appears to be baked into existing theories of causation.

8.2.1. The Argument from Intuition

As we see it, there are at least three ways to formulate the argument from intuition. The first of these is a 'swapping' argument based on the consideration of canonical cases of causation. To see the idea, consider the canonical example of a bottle smashing after a ball has been thrown at it. In such a case, it's clear that the throwing of the ball is the cause of the smashing of the bottle. Intuitively, however, there must be a temporal betweenness relation of the kind that we find in a C-series, whereby we can say that the striking of the bottle by the ball is temporally between the throwing and the smashing. We can see this because, if we try to swap in some other kind of relation to scaffold the case, things go wildly wrong.

For instance, if we consult our intuitions about a case where we *spatially but not temporally* separate the throwing from the smashing, then we don't recover a case of causation. Suppose that we have, at t_1, two spatially separated events: e_1 and e_2; e_1 is the throwing of a ball; e_2 is the smashing of a bottle. Let's allow that e_1 is located 6 feet away from e_2. We have the same basic structure that we had in the bottle smashing case, except we've swapped out any and all temporal relations between e_1 and e_2 for spatial relations, and we've specified that there's no temporal distance between e_1 and e_2. What we find is that, where we swap out temporal relations between e_1 and e_2 for spatial relations, the causal relation disappears too. What's more, for every case of causation we find that matters are the same.

We are willing to concede the intuition at play in this kind of case. Intuitively, swapping temporal ordering relations of the kind found in a C-series, for spatial relations, in some cases of causation does undermine the causal structure of the case. But we think that, at best, what this shows us is that some familiar instances of causation require a C-series. It does not follow from this that causation in general necessitates the existence of a C-series.

To see the point a bit more clearly, it is useful to consider an analogous case. For instance, consider intuitions about the composition of material objects. Suppose we have a specific instance of composition such that all parts are spatially related to one another. A solar system is a nice case in point; each planet is a part of the system, and each planet is spatially related to every other. Suppose, then, that we swap the spatial relations for temporal relations, such that the planets are no longer spatially related to one another but *are* temporally related to one another. In such a case, intuitively, at no point in (or indeed across) time is there a composite object that can be identified as the solar system. It would seem, then, that swapping the spatial relations for temporal relations has made reality such that the planets are no longer a part of the solar system. Well and good. But this does not show that composition cannot obtain across times. Rather, it shows that this *instance* of composition does not obtain across times in the absence of spatial relations connecting the parts at a given time.[1] That, of course, is quite a different thing.

And, predictably, things will work in the other direction, too. Suppose that we adopt a temporal parts view of persistence, such that there are composite objects that span times, and we consider what it would take for persons to persist over time. On this view, each persisting person is made up of temporal parts that are connected to one another in just the right kind of way—let's suppose that psychological continuity *is* that 'right kind of way'. Now consider Freddie. Freddie is identical to the collection of temporal parts that are connected to one another by both temporal relations *and* relations of psychological continuity. If we 'swap' those temporal relations for spatial relations, it's clear enough (we think) that we no longer have a composite object that is identical to Freddie that is spread across multiple spaces. What we do have, we're not quite sure—but we are sure it isn't Freddie—it looks to be something of a mess. So, in this case, too, by swapping temporal relations for spatial

---

[1] Either because there is no composite thus composed of those planets, or because there is, but the resulting composite is not a solar system.

relations we've removed the existence of a particular object. But, of course, that doesn't show that composite objects can't span spaces. At best, it shows that persons don't span such spaces.

The moral of all of this? What our opponent is insisting on is that there are strong intuitions about causation: causation can only occur in the presence of temporal ordering, and we can see this once we realise how obvious it is that we don't have causation when we swap relations of temporal separation for spatial separation in a particular case. Our point in reply is simply that this strategy is ineffective. Showing that an instance of a relation R *does not* hold in the absence of some background condition, C, fails to show that R *cannot* hold in the absence of C.

As noted, there are at least three ways to formulate the argument from intuition. The second form of the argument does not involve reflecting on cases. Instead, the thought is that when we reflect on the notion of causation we have a powerful intuition to the effect that causation requires the temporal separation of cause and effect. Temporal separation, so the thought goes, requires the existence of a temporal ordering, minimally: a C-series.

Our response to this second argument from intuition is to invoke a controversial topic in discussions of causation: simultaneous causation. We will describe two kinds of case involving a cause and effect that stand at no temporal distance from one another. We'll do so in order to suggest that, once we start to reflect on cases, it's less clear than one might suppose that we need a relation of temporal separation (of the kind generated by a C-series) that connects cause to effect.

The first kind of case is very familiar from the literature on simultaneous causation. Consider the case of a cannon ball resting on a cushion. The cannon ball resting on the cushion, at t, is the cause of the indentation in the cushion, at t. Or, consider the case of two books on a shelf, leaning against one another. The cause of the book on the left staying upright at t_1 is the book on the right leaning against it at t_1.

Of course, if such cases were genuine cases of causation, then this would show that we can have causation in cases where there is no temporal separation between cause and effect. To be sure, it wouldn't yet show that we can have causation in the absence of a C-series, but since the second argument from intuition depends upon a notion of temporal separation, it would do a good deal to help our cause. So, what should one say about simultaneous causation?

We suspect that responses here will vary. Some recent work has looked to make the case for simultaneous causation (for instance, Huemer and Kovitz,

2003; Mumford and Anjum, 2011: 111). Others have argued against it (for instance, Le Poidevin, 1991: Chapter 6). We don't propose to add to the literature here. Our goal is more modest. Simply: anyone open to the mere *possibility* of simultaneous causation should not take the second argument from intuition to be decisive.

This brings us to the second kind of case. This case *is* slightly exotic, but we think it's straightforward enough. Elizabeth is an inventor. For several years, she has been working on a time machine. She finally thinks that she has cracked it. Wary of paradoxes (banana skins, and other potential threats to her person), to test the machine, Elizabeth decides to send the machine away from her in time and space and automate its return. She sets up her experiment as follows. In her Lab, L_1, she will, at t_1, send the machine into the future, t_2, to wait there until t_3. She automates the machine to then return to t_1. To avoid the machine damaging anything in her lab, she decides to automate the return to the lab next door, L_2. Between L_1 and L_2 there is a viewing window. To provide some visual clue to her success (or otherwise), Elizabeth sets the machine up to emit a flash upon its return to t_1 when it reaches L_2. Lo and behold, when Elizabeth flicks the switch the machine disappears from L_1 and there is a sudden flash of light from L_2 (later checking of recording apparatus will reveal that the machine arrived in L_2 at t_1). Elizabeth waits in her lab until t_2 at which point the machine appears in her lab, L_1, and then, at t_3, the machine disappears.

Crucial for our purposes is the connection between Elizabeth's flicking the switch to initiate the automated processes, and the flash of light emitted from the time machine in the lab next door. In this case it seems reasonable to suppose that the cause of the time machine emitting a flash of light at (t_2, L_2) is Elizabeth's flicking the switch at (t_1, L_1). After all, it is Elizabeth's action in her Lab at t_1 that causes the machine to travel to the future, and then travel back to her present, emitting a flash of light when it does. Even if we don't think that causation is a transitive relation, it is clear enough that some instances of causation are such that x causes y, y causes z and x thereby causes z. This seems, to us, to be one such case. The cause of the time machine emitting a flash of light at (t_1, L_2) is Elizabeth's flicking the switch at (t_1, L_1). A consequence of this is that the cause is simultaneous with its effect. What this shows, then, is that we do not need a temporal gap between cause and effect.

There are all sorts of response we can imagine an opponent making. Perhaps they will object that there *is* still a temporal gap between cause and effect—it's just that the gap consists, not in objective time, but in the 'personal time' of the time machine. Perhaps they will object that time travel in general is not possible. Or perhaps they will object that some specific detail of the case is problematic.

Whilst we think that there are sensible replies to these sorts of objections, in any case we think that they miss the point somewhat. The claim that we're pursuing here is that it's supposed to be intuitive that causation requires a relation of temporal separation in virtue of which cause and effect can be ordered in time. What we take ourselves to have shown, however, is that there are cases where this is not so—where, at least intuitively, there is nothing problematic about there being a scenario in which there is no such temporal separation. If that's right, then we don't see that there is in fact a clear intuition that causation occurs only at a temporal distance. Rather, we take causation without temporal separation to be surprising—we take it to be the kind of phenomena with which people are generally unfamiliar. Nonetheless, we don't think it's apparent that causation without temporal separation is impossible. We certainly don't think it's obvious that *all* instances of causation require temporal separation between cause and effect, even for ordinary causal happenings.

We come now to the third and final argument from intuition. This third argument is the most direct of the three. If we simply reflect on the nature of causation, so the thought goes, then we see that the reality of causation requires the reality of (at least) a C-series. While we agree that causation in the absence of a C-series is odd, we don't think much of this third appeal to intuition. This is not because we take a particularly sophisticated meta-philosophical position on the role of intuitions in philosophy. Indeed, we're prepared to allow that intuition *might* do quite a lot of work in philosophical inquiry. But we do think that there is a useful distinction to be drawn between something's running counter to our intuitions, and something being a bit surprising because we've not properly thought about it before. The trouble is that it can be very difficult to tell, from introspection alone, whether something is counterintuitive in the manner that might provide evidence against a philosophical thesis rather than simply being surprising because it is unfamiliar. Moreover, when the thesis itself is unusual—as it surely is in the case at hand—it can be doubly hard to differentiate between the counterintuitive and the surprising.

Of course, sometimes 'counterintuitive' simply means 'incoherent', and when understood in this way it is easier to differentiate things that are counterintuitive from things that are merely surprising, since only the former will be incoherent. Note, however, that 'incoherent' in this context usually means 'inconsistent'. There is nothing inconsistent, so far as we can tell, about the idea that causation occurs without a C-series. We can see this simply by considering that there is nothing incoherent about the idea of causation between things that are purely spatially separated from one another. To be sure, one might have the view that causation of this kind is metaphysically impossible (because ruled out by one's favoured theory of causation) but it doesn't follow from that, at least not without substantial argument, that the imagined scenario is incoherent.

Now, one could take this last point on board and argue that 'counter-intuitive' means something like 'it seems metaphysically impossible'. But what seems metaphysically impossible is typically informed by one's back-ground theories about metaphysics. In this way, 'x seems metaphysically impossible' just starts to look like a way of forcing a particular theory of causation, namely one that requires the C-series. To be sure, and as we discuss below, there are such theories. But as we shall also see, this does not do much to show that causation requires the existence of a C-series. One might disagree with this: metaphysical seemings are just primitive intuitions. Perhaps they are. But then we return to our earlier point: we are unsure how to differentiate these metaphysical seemings from the surprise one might enjoy when encoun-tering an unfamiliar idea.

No doubt there are other ways to develop the argument from intuition. Our aim is not to be exhaustive in this respect. We hope to have said enough, however, to show why we don't find these kinds of arguments compelling. Either we don't know what to make of the intuitions or we can find intuitions based on other cases that pull in a different direction. We don't think this is particularly surprising. Intuitions are rarely decisive on their own. They typically need to be supplemented by arguments. One option might be to supplement the appeal to intuition with arguments based on the examination of specific theories of causation. It is to this type of approach that we will now turn.

8.2.2. The Argument from Theory

This brings us to the second broad type of argument for the necessitation of the C-series by causation. This type of argument arises in view of the fact that theories of (rather than intuitions about) causation are often fundamentally temporal. As analyses of what *causation is*, they operate with the assumption that temporal ordering of the kind found in a C-series is real. They fail if temporal ordering of the relevant kind is not real. And so, to the extent that an opponent is attracted to any extant theory of causation, they should reject our position.

To make the point maximally vivid, consider the process theories of causation, of the likes defended by Salmon and Dowe. Process theories of causation begin with the observation that, at least typically, where we have causation we have a flow of energy (cf. Fair (1979)). From that starting point, different analyses of causation can be provided. For instance, Dowe (1992: 200) summarizes Salmon's position as follows:

(I)   A process is something which displays consistency of characteristics.

(II) A causal process is a process which can transmit a mark.

(III) A mark is transmitted over an interval when it appears at each spacetime point of that interval, in the absence of interactions.

(IV) A mark is an alteration to a characteristic, introduced by a single local interaction.

(V)   An interaction is an intersection of two processes.

(VI) A causal interaction is an interaction where both processes are marked.

It's easy enough to see how Dowe's analysis of causation is problematic if there is no C-series. A process is, intrinsically, something temporal— something that cannot exist without at least the minimal kind of temporal ordering afforded by a C-series. Moreover, condition (III) explicitly requires the existence of space-time points, and, as previously discussed, spacetime structure is likely sufficient for the existence of a C-series. And, last, 'interactions', like processes, are intrinsically temporal, requiring at least a minimal sort of temporal ordering. If there is no C-series, then it's very hard to see how we can recover causation of this sort.

Whilst the point is maximally vivid with respect to something like a process theory of causation, it's also easy enough to show how the same basic problem will arise for other, more general accounts—of both a Humean and Aristotelian stripe.

Consider, for instance, a standard counterfactual analysis of causation of the kind defended by Lewis (1986). Let 'c' and 'e' stand for particular events, and '$O_{(c)}$' and '$O_{(e)}$' stand for 'c obtains' and 'e obtains', respectively. Then, e is causally dependent upon c iff:

$$O_{(c)} \square \rightarrow O_{(e)} \text{ and} \sim O_{(c)} \square \rightarrow \sim O_{(e)} \text{ (p.167)}$$

In other words:

> A counterfactual "If it were that A, then it would be that C" is (non-vacuously) true if and only if some (accessible) world where both A and C are true is more similar to our actual world, over all, than is any world where A is true but C is false' (p. 41)

And, as is familiar, we can give the similarity metric as follows.

1. It is of the first importance to avoid big, widespread, diverse violations of law.
2. It is of the second importance to maximize the spatiotemporal region throughout which perfect match of particular fact prevails.
3. It is of the third importance to avoid even small, localized, simple violations of law.
4. It is of little or no importance to secure approximate similarity of particular fact, even in matters that concern us greatly (Lewis 1979: 472).

Lewis's similarity metric requires us to consider spatiotemporal regions. The idea underpinning this formula is that in evaluating a counterfactual, we first match worlds based on laws and history leading up to the point at which a given event x occurs. Those worlds that have the same history until x occurs, and that have similar laws, are the closest worlds. Of these so-far-closest worlds, we focus on those in which x does not occur at the relevant point and look to see what happens in the short period afterwards. If the closest world in which x does not occur is also a world in which y does not occur, then the counterfactual is true. Otherwise, the counterfactual is false.

Now, of course, both the formula and the explication we have given make explicit reference to things like 'spatiotemporal region' and one event 'being followed' by another. Lewis's analysis thus seems to presuppose the existence of spacetime and, with it, structure enough for the existence of the C-series.[2] Perhaps there is a way to modify the theory (more on this later) but, at least in its bare form, the counterfactual analysis of causation seems to imply the existence of a C-series by committing us to the existence of spacetime.

Similar thoughts apply to a neo-Aristotelian model of causation, underpinned by dispositions. With some variation, a position sometimes adopted is

---

[2] A further wrinkle emerges once we take into account that Lewis thinks that the direction of counterfactual dependence gives us the direction of time. We set this to one side.

one according to which causation is underpinned by the exercise of dispositional causal powers. Illustration: the property of solubility is naturally thought of as a disposition to dissolve in water, and so placing the soluble material into water is often taken to be the cause of its dissolving. This dispositional nature is not typically taken to be reducible by those who hold such views. Focus attention on these powers, however, and let us stay with the case of the disposition to dissolve in water. This property—this disposition—is directed. If a soluble object is placed into water, then what that property does is direct the object to a state of being dissolved *at some future time*. Thus, the property itself looks to presuppose temporal ordering in some minimal sense, and so would seem to require at least a C-series (this being the most minimal temporal ordering that we can think of).

Furthermore, any analysis of causation that is predicated on powers, will, so far as we can tell, need to include reference to both stimulus conditions and manifestation conditions. To explain: consider the case of a soluble medicine. The stimulus condition for the property of solubility is the condition of being placed in water. The manifestation condition for the property is the condition of being dissolved. If we attempt to give an analysis of causation in terms of dispositions, then we will want to say that there is a temporal asymmetry here: that in order for one event to be the cause of the other, we must first appeal to the stimulus condition and *then* to the manifestation condition. To give just a bit more detail: we might be able to say that the placing of the medicine in water is the cause of it dissolving, because the medicine was first placed into the stimulus condition for its property of solubility, which led to the manifestation condition for the property of solubility. As with dispositions, manifestation and stimulus conditions seem to require some minimal notion of temporal ordering.

We don't at all mean to suggest that this little overview is exhaustive. It is only intended to be illustrative. The point is just that theories of causation are typically set up (as are those above) in such a way as to first identify two (or more) putative causal relata that are temporally ordered with respect to one another in at least a C-series. From there, the analyses will seek to provide a reductive account of how to understand the nature of a causal relation that connects those temporally separated causal relata. Whether we understand that relation in terms of processes, counterfactuals or dispositions, as we have considered here, the basic structure of the analyses is always such that it sets out to connect temporally separated relata. As soon as we set off down that road, we're on a hiding to nothing in a world that lacks a C-series. And, to return to the theme at the start of the sub-section, anyone attracted to any extant theory of causation would thus seem to have grounds to reject the claim that there can be causation without something answering to the folk conception of time.

Ultimately, we don't think that the argument from theory is compelling. The core of our response is very simple. Extant models of causation treat it as a phenomenon that requires temporal ordering and thus, minimally, a C-series. That being so, it's entirely unsurprising that extant models end up being incompatible with the absence of a C-series. Now, if there were some broad consensus in metaphysics that one of the models we have described is the correct account of causation, then we would have reason to worry. But there simply is no such consensus. And we know of no argument, beyond the raw appeal to intuition outlined above, for the claim that causation must involve a C-series, and thus that an acceptable model of causation presupposes the folk notion of time.

One might argue that the fact that so many theories of causation converge on the need for a C-series despite being otherwise diverse does provide some basis for the view that the C-series is necessary for causation. We don't find this line of thought to be particularly compelling, however. Just because most theories of a certain phenomenon make a particular assumption, it does not follow that we thereby have good evidence for the assumption merely in virtue of such convergence. At best, we have whatever evidence for the assumption is given to us by evidence for the theories. We don't get an evidential boost from agreement between metaphysical theories (unless we have reason to suppose that they are all correct in a certain respect). And, in any case, that there is convergence here is unsurprising. It is a background assumption of all models of causation that time is real. As such, the various theories of causation look to model causation *as* a timely phenomenon. The question we should be asking then, is not 'are current models of causation compatible with the absence of a C-series?" Rather, we should be asking "can we model causation in a timeless world?' That is, is it possible to provide a model of causation that does not presuppose a C-series ordering?

How do we do that? Well, we aren't going to try to simply import a theory of causation that assumes the existence of a C-series to a world where time isn't real. What we can try to do, however, is to transport some of the relevant features of one or more of these theories into a timeless setting. That is what we aim to do in the next section.

8.3. Approximate Spacetime and Causation

Recall Lewis's counterfactual theory of causation. Abstracting away from the details a bit, the theory has three parts. First, the theory analyses causation in terms of counterfactual dependence. Second, the theory focuses on particular events as the relata of the causal relation, and thus it is propositions about those events that we find in the subjunctives that form the backbone of the theory. Third, the theory makes use of a particular semantics for counterfactuals that deploys a closeness measure over possible worlds, one that is defined by the four-step recipe noted above.

What resources do we have available for transforming Lewis's counterfactual theory into something that is fit for service in a timeless setting?

When addressing the problems of empirical coherence that arise for various approaches to QG, we appealed to the idea that something approximating GR spacetime exists, even though GR spacetime itself does not. In particular, at low-energies, the physical structure at issue—the one posited by quantum gravity—can be approximately described using GR without getting into trouble. That is not to say that GR is a *true* description (after all, the entity being so-described is not spacetime), but GR remains an empirically adequate theory, in the low-energy regime.

In short, while we don't have spacetime we do have an entity that is approximately spatiotemporal, i.e., an *approximate spacetime*. Our proposal for rebuilding Lewis's counterfactual theory of causation is rather simple: find and replace 'spacetime' with 'approximate spacetime' within the theory. Because most of our causal judgements are made with respect to the low-energy regime in which approximate spacetime exists, the substitution should not be all that disruptive. Of course, as we shall see, things are bit more complex than that. But that, at least, is the basic idea.

Let us start with Lewis's semantics for counterfactuals. Recall that, according to Lewis, a counterfactual of the form *if A had been the case, B would have been the case*, is true if some A world in which B is true is closer than any A world in which B is false. As we saw, the key to the semantics is a closeness measure over worlds. Spacetime is used to anchor the closeness measure in two places: first, in terms of the particular ordering used to define the measure and, second, in terms of how worlds, for Lewis, are defined.

Take the definition of worlds first, since it is the easier to handle. For Lewis, a world is defined as a maximal, complete concrete spacetime. This won't do for our purposes, since it would have the unwelcome result that the actual world is not a possible world (assuming, as per the discussion in Chapters Six and Seven, that spacetime does not emerge and thus does not exist). Apart from anything else, this will cause problems for the closeness measure, since we will be comparing the closeness of possible worlds to some impossible (actual) world, rather than possible worlds to some possible (actual) world. We can address this issue by adopting an ersatz account of worlds, thereby treating worlds as sets of sentences in a world-building language (e.g., Lagadonian sentences). A possible world is a set of such sentences that is maximal and complete. Truth in a world is just membership. Thus, a proposition is true at a world if it is expressed by a sentence that is a member of the set of sentences that constitute that world. Worlds thought of this way need not be spatiotemporal.[3]

With respect to the closeness measure itself, we propose to swap spacetime for approximate spacetime, yielding a more-or-less unchanged version of the Lewisian metric along the following lines:

1. It is of the first importance to avoid big, widespread, diverse violations of law.
2. It is of the second importance to maximize the **approximately spatiotemporal** region throughout which the perfect match of particular fact prevails.
3. It is of the third importance to avoid even small, localized, simple violations of law.
4. It is of little or no importance to secure approximate similarity of particular fact, even in matters that concern us greatly.

So, for instance, suppose we adopt the broad picture recommended by loop quantum gravity (LQG). As discussed in Chapter Five, on this view the fundamental physical structure is represented by a spin-network. Some spin-networks take the form of weave states, such that at low energies, the mathematical description of the spin-network structure achieves numerical agreement with the spacetime metric of GR. A spin-network at low energies is one candidate for being approximate spacetime, and it is what we will use to work through the details of the closeness metric.

In fact, because of the way that the spin-network structure approximates spacetime, and because we can use GR as an approximately adequate

---

[3] Of course there will be a set of worlds that are spatiotemporal in the following sense: the proposition 'spacetime exists' will be expressed by some member of every world in the set.

description of the physics at low energies, the way in which we work through the evaluation of a counterfactual need not change much at all. So long as we are focusing on counterfactuals involving entities within the low energy regime, we can use the approximate spatiotemporal description of a weave state without getting into trouble. This means that we can evaluate counterfactuals *as if* spacetime existed, given the close agreement between approximate spacetime and GR spacetime.

To see this in a bit more detail, let us work through the metric line-by-line. We start with an approximate spatiotemporal description of the low-energy regime of the actual world. This gives us our baseline for comparing worlds for closeness with respect to actuality. We then consider all worlds that match the actual world as closely as possible with respect to the laws of nature. These will, at a minimum, all be worlds in which the low-energy regime can be described using GR. We then order worlds in terms of how closely the GR descriptions of the low-energy regime match. This is how we understand closeness of match between approximate spatiotemporal regions outlined in the second line of the metric stated above. We then order worlds in terms of any small differences in the GR description, and so on until we reach our final ordering. On our picture, every counterfactual about something putatively located in GR spacetime is made true by modal facts concerning something located in a spin-network, under an approximate spatiotemporal description.

One choice we must make here concerns the types of nomic violation that we should pay attention to. So, for instance, compare two worlds: a world in which the underlying physics is correctly described by LQG and another world in which the underlying physics is correctly described by a different approach to QG altogether; causal set theory, say. Suppose that the two worlds admit of quite different low-energy descriptions, only one of which resembles the actual world (the causal set description, say), though both are captured by GR in a broad sense (perhaps, in the low-energy regime, the two worlds correspond to different solutions to the field equations). Finally, suppose that the actual world is a world in which LQG is true. Which world is closer to actuality: the LQG world with a vastly different low-energy description but a correct high-energy description, or the causal set world that matches the low-energy description of the actual world but gets the high-energy physics wrong?

Here we must make a choice about what we value more: do we value similarity at the low-energy level more than similarity at the high-energy level or vice versa? Whether the choice matters depends on two things. First, on the kinds of counterfactuals we are assessing and, second, on how sensitive the facts in the low energy regime are to the high-energy physics. As noted, the counterfactuals we are assessing are ones concerning goings-on in the low-energy regime, since this is the regime in which empirical confirmation occurs. If the low energy physics is invariant with respect to the high energy physics—i.e., changes to the physics of the high-energy don't make a difference to the low-energy description—then we can simply match worlds in terms of their low-energy descriptions. If, however, the low-energy description is sensitive to changes in the high-energy physics, then we should order worlds with respect to similarity in both the high- and low-energy regimes. Staying as close as we can to Lewis's original proposal, we recommend matching the high-energy regime first, and then matching for the GR description. Indeed, doing so is more or less demanded by the four-step metric. While GR is empirically adequate, it is not true, and so doesn't qualify for lawhood in even the actual world. The high-energy physics, by contrast, does constitute a lawful description. So, the first line of Lewis's recipe tells us to match at the high-energy level first, and then carry out any matching at the low-energy level, approximately described by GR.

The central difference between Lewis's account of counterfactuals and our own pertains to what, exactly, we are matching in the second line. In Lewis's original formulation, we match spatiotemporal regions. In our formulation, we match regions that can be approximately described using GR. The two systems agree, however, in that the way to match either spatiotemporal regions or approximately spatiotemporal regions is to compare the GR descriptions and use that as the anchor for similarity. The upshot is that *for a certain set of counterfactuals* the two accounts will yield the same truth-values for those counterfactuals. In particular, for any counterfactual in which both the antecedent and the consequent pertain to facts that live entirely within the low-energy regime, the two systems agree. Because all of the day-to-day counterfactuals that we employ involve objects within this regime, all of our ordinary counterfactuals will receive the same treatment as under Lewis's account. Most importantly, for our purposes, since all of the empirical observations we carry out are at low energies, all of the counterfactuals implicated in empirical confirmation will take the truth-values that we expect.

One might demur: the counterfactuals are not *exactly* the same. Granted, the same method—broadly construed—can be used to evaluate counterfactuals that are very similar to the ones that Lewis considers. But the counterfactuals themselves are different enough that we cannot use them to support a viable account of causation. One way to see the worry is in terms of Lewis's notion of an event. According to Lewis's account, it is not the case that just *any* instance of counterfactual dependence is sufficient for causation. Lewis recognises that there are many true counterfactuals that should not be interpreted in causal terms. For Lewis, a counterfactual implies causation only if it involves propositions describing distinct *events*. An event is, in turn, analysed in spatiotemporal terms: an event, for Lewis, is an object with a particular property, at a particular spatiotemporal location. If events are essentially spatiotemporal, or essentially temporal, then there simply are no events without spacetime or space and time. Since the view we are developing is premised on the idea GR spacetime does not exist and nor does a C-series, one might contend that such a view thereby implies that there are no events. Thus, while one might be able to countenance counterfactual dependence between certain 'bits' of one's ontology, these relations of counterfactual dependence cannot count as causal relations because they do not link distinct events.

Recall, however, that while we do not have spatiotemporally located events, we do have entities located in some physical structure that can, under certain conditions, be described as approximately spatiotemporal. The counterfactuals to which we have appealed, link these approximately spatiotemporal entities. To be clear, though, the counterfactuals themselves just speak of the usual menagerie of entities: doors, tables, cats, people and so on. It is just that *what these entities are* is not to be understood in spatiotemporal terms, and so while the same counterfactuals may be true (i.e., counterfactuals that mention doors, people, cats and so on), the entities that the counterfactuals describe are not to be understood in the usual way. Now, maybe counterfactuals of this stripe are not sufficient for causation in Lewis's strictly spatiotemporal sense, but we believe that such counterfactual dependence is sufficient for causation in *some sense*.

We think this because the counterfactual dependencies at issue play the same role as the ones that Lewis deems to be causal. The counterfactuals can be used for *prediction*, at least within the low-energy regime. Because the underlying nomic framework for the counterfactuals at issue is the approximate GR description, and because this description is empirically adequate, we can make predictions in terms of what would happen under various counterfactual suppositions. Those counterfactuals will pan out as expected because they are underpinned by a physical description that works, even if it is not strictly speaking true.

In addition, and as already indicated, the counterfactuals can underwrite *observation*. Suppose that Sara observes a cat walking past her door. We can analyse this instance of observation in terms of a counterfactual: if the cat had not walked past the door, Sara would not have seen it. This counterfactual is true in virtue of the fact that there is an approximately spatiotemporal event in which an entity that approximates a spatiotemporal cat can be approximately described as moving in the way described by GR.

Finally, the counterfactuals at issue can meaningfully underwrite *manipulation and control*. Suppose Sara wants to open the door. To do that, she needs to turn the knob and pull. This instance of manipulation is underpinned by a counterfactual: if Sara were to turn the knob on the door and pull, then the door would open. This counterfactual remains true even without GR spacetime. The only difference is that when we evaluate this counterfactual we match worlds in terms of approximate spacetime, rather than GR spacetime, and the counterfactual itself links an approximately spatiotemporal entity—the handle and the turning—with another approximately spatiotemporal entity—the door opening.

Perhaps there are other roles for causation than the three considered. But for those roles, too, we believe that our relation of counterfactual dependence is at least as good as the one on which Lewis focuses. Indeed, for any job description that one might have for causation, we can be confident that our account will be able to play that role. Moreover, notice that our account does not hang on any particular account of what's going on at the level of high-energy physics. So long as the high-energy physical theories reduce to GR at the level of description where we want to make causal claims, we see no reason why causation cannot be recaptured.

Of course, matters are different for the high-energy level. There, it is not clear what to say about causation. But, by the same token, we take it that there is less pressure to recover a viable notion of causation at high-energies, given that all of the empirical observations that we have, and deem ourselves broadly capable of making, happen at low energies. Still, it might be thought that some more fundamental picture of causation is needed to make sense of the metaphysics of causation we are proposing. Whether that is correct is an issue we return to in the next section.

8.3.1. Objections

So far, we have sketched out how to recover one account of causation using an entity that approximates spacetime, rather than GR spacetime itself: a counterfactual theory. Shortly, we will suggest a way of generalising our recovery strategy to other metaphysical accounts of causation. Before doing so, however, and in order to frame the discussion of such a generalisation, it is useful to address a couple of objections to the account developed so far.

First, on our view, approximate spacetime plays the same role with respect to supporting causation as does spacetime proper. But, of course, there are differences between spacetime and an approximately spatiotemporal structure. These differences, one might argue, undermine the capacity for approximate spacetime to support causation. One way to press this point is to argue that precisely what we lose when we shift from spacetime to approximate spacetime is spatial and temporal structure. But (at least) time, in the sense of a C-series, is necessary for causation and so approximate spacetime simply cannot underwrite an appropriately causal metaphysics.

Whilst this is not a compelling line of argument (as we have already argued in the first half of this chapter there is no reason to suppose that the loss of time entails the loss of causation) a more promising way to make the objection is to focus on the particular differences between spacetime and approximate spacetime found within specific approaches to QG and to argue, based on these differences, that approximate spacetime cannot do the work we are asking of it. There are, as indicated in Chapter Five, many approaches to QG available, and obviously we cannot provide a complete catalogue of the differences between spacetime and something that might be called approximate spacetime in these theories (insofar as there is such a thing). What we can do, however, is consider the main lines of difference and speak to the role that these differences might play in undermining the existence of causation. After that, we'll offer a general reason why one should not be too worried by the style of objection at hand.

A brief survey of the various approaches to QG reveals at least four differences between spacetime and approximate spacetime. First, spacetime is continuous whereas the physical structure that approximates spacetime is discrete—according to at least LQG and causal set theory. Second, in some approaches to QG the underlying physical structure has quantum properties (such as existing in a superposition, or obeying the Wheeler DeWitt equation) whereas spacetime is purely classical. Third, spacetime obeys the Lorentz symmetries whereas some physicists predict the breakdown of Lorentz invariance at the high-energy level. This is, for instance, predicted for causal set theory. Fourth, spacetime is a purely local theory, in the sense that any effects propagate to nearby locations at a velocity that is capped by the speed of light. The high-energy physics, by contrast, may be non-local. Worse, purely local relationships in the high-energy description may correspond to non-local relationships in the approximately spatiotemporal description (this is the issue of disordered locality for LQG touched on briefly in Chapter Five, more on this in a moment).

We have taken seriously the presence of causation in theories with at least some of these features in the past. Newtonian mechanics is not Lorentz invariant, and yet there is no temptation to say that it therefore cannot be causal. Similarly, quantum mechanics displays many of the same peculiar quantum features as do some approaches to QG (such as superpositions and entanglement) and here, again, there is little temptation to think that quantum theories in general are unable to handle causation. Discreteness also doesn't seem to pose a substantial problem, particularly once it is noted that most of our ordinary causal thinking involves discrete events. The only effect that discreteness would have on causation is to make causes and effects discrete, but that is already a tacit (though perhaps ideal) assumption of many causal theories; Lewis's theory is a fine example.

The difference that is perhaps most important, for our purposes, is the difference in locality. There is a long tradition of denying that causation can be non-local, and so if the only relations of counterfactual dependence, say, that we can recover within the low-energy regime are counterfactuals connecting entities that, according to the approximate GR description being used, are non-local, then that would be a problem. Consider the problem of disordered locality for LQG. In LQG entities that are, say, adjacent in a spin-network at high-energies, may be arbitrarily far away from one another when described using GR within the low-energy regime. This might seem to pose a problem for our account of causation (cf. Huggett and Wüthrich (2013)).

However, we are only forced to countenance causal relations between such entities that are non-local by the lights of the GR description if we allow that an instance of causation occurs between these entities at the high-energy regime, and that this instance of causation is preserved into the low-energy regime. And we don't see why this should be so. Indeed, the very fact that GR can be used to approximately describe the low-energy physics without getting into trouble would seem to forbid the transmission of non-local causation from the high-energy description to the low-energy description. After all, a central component of the approximate GR description is that any causal relations that obtain are purely local. It would seem then that any non-local causal relations at the high-energy level must be broken at lower energy scales.

This last point helps us to outline a more general response to the objection under consideration. The objection has it that the differences between approximate spacetime and spacetime undermine the existence of causation. But, for our purposes, the similarities are much more important than the differences, since it is the similarities between spacetime and approximate spacetime that ground our account of causation. So long as we have an object that we can approximately describe using GR, we can rebuild an account of causation that matches the account we have if spacetime exists. Even if the object that behaves as though it were spacetime is different from spacetime itself, it is the fact that it behaves in a spatiotemporal way that makes causation possible. Whatever the differences might be with spacetime itself, these just don't matter for causation.

This brings us to the second objection against our account of causation. It is a central part of our view that approximate spacetime provides the foundations for causal structure. But, as we have also made clear, the physical structure posited by a given approach to QG only behaves as though it were spatiotemporal under specific conditions. At high energies, the physical structure at issue is not even approximately spatiotemporal. So, there is, one might think, little hope of recovering a viable notion of causation within the high-energy regime. Thus, causation, on our view, is *emergent*. But now a metaphysics of emergent causation is needed, and one might worry that this is going to require an entirely new metaphysical investigation. For just as it can be *prima facie* difficult to see how spacetime might emerge from a fundamentally non-spatiotemporal structure, one might argue that it is similarly mysterious to see how causation might emerge. Of course, matters would be different if spacetime were also emergent. Then there would be no problem with emergent causation. But on our view spacetime doesn't emerge, and yet it looks like causation does.

Since we don't want to provide a metaphysics of emergent causation, that is not the strategy that we will pursue here. Instead, we offer two ways forward. The first, and most ambitious, is to extend the counterfactual theory of causation even further, into the high-energy regime, where not even approximate spacetime exists. The second is to deny that causation exists—as characterised by a counterfactual theory, or indeed any of the standard theories of causation that we find in metaphysics. Indeed, we deny that causation, so understood, exists at either low energies or high energies. What exists, rather, is a relation that is *approximately causal* at low energies, but that is not approximately causal at high energies. We take each suggestion in turn.

8.3.1.1. Causation in High-Energy Regimes

Providing an account of causation in high-energy regimes requires departing significantly from Lewis's strategy for evaluating counterfactuals discussed in the previous section. No simple 'find and replace' solution will work. Still, applying Lewis's approach to high-energy, non-spatiotemporal physics is not as crazy as it might first sound. Counterfactual dependence does not *require* spacetime. Methods for meaningfully evaluating counterfactuals in the absence of spacetime have been offered for pure mathematical cases, metaphysical cases and logical cases (see, e.g., Baron et al. (2017, 2020) and Schaffer (2016)). It turns out that all we really need for counterfactuals is a bare minimum of two things. We need a physical structure in which different locations can be individuated, and we need rules over the structure that dictate how what is true at one location determines, or otherwise has implications for, what is true at another location. Given this structure, it is then possible to evaluate counterfactuals in a meaningful way.

The existence of the fundamental physical structures posited by some theories of QG satisfy these two bare conditions for counterfactual dependence. Consider, for instance, LQG with its spin-networks. Both the nodes and edges of spin-networks can correspond to locations: they are unique, in the way that locations are, and they can be individuated from other nodes and edges, in the way that locations generally need to be. The theory itself provides rules that, in principle at least, allow us to draw inferences about what is true at one location in a spin-network based on what is true at another. What we can do, then, is modify Lewis's metric for counterfactuals by appealing not to *approximate spacetime* but, rather, the more fundamental spin-network structure. A metric along these lines uses 1, 3 and 4 from Lewis's metric and replaces 2 with:

2. It is of the second importance to maximize the **spin-network region** throughout which perfect match of particular fact prevails.

When we are considering the low-energy regime, the metric collapses into the metric stated above, involving approximate spacetime. Spin-networks in a weave state behave as if they are spatiotemporal at low-energies, and so regions of spin-networks correspond, broadly speaking, to regions of approximate spacetime. In the high-energy regime, however, the metric still applies, only here it is not applying to regions that are behaving in a spatiotemporal manner. Rather, the metric applies to regions of the physical structure represented by spin-networks. This metric thereby permits the evaluation of counterfactuals involving entities located within the broader spin-network structure.

We have put the point in terms of spin-networks and weave states but, of course, it generalises beyond this. Whatever the fundamental physical structure might be, so long as it defines a notion of location (which, we are assuming, it does) it will be possible to evaluate counterfactuals in much the same way. This leads us toward a fully general version of the metric, broadly applicable to any approach to QG that employs a non-spatiotemporal notion of location. This version of the metric replaces approximate spacetime with a notion of physical region, which leaves unspecified what this might be, leaving it to physics to fill in the details:

2. It is of the second importance to maximize the **physical region** throughout which perfect match of particular fact prevails.

The first response to the worry about emergent causation, then, is to just shift into this more general framework and use an account of causation for both high- and low-energy regimes. As noted, however, there are two responses available to this concern.

8.3.1.2. Causation vs Approximate Causation

The second response is to deny that there is causation at all. In its place, we can posit a relation that is *approximately causal* at low energies, but that is not approximately causal at high energies. We can use a view like Lewis's to describe this causation-like relation at low energies without getting into trouble (at least, no more trouble than we already get into by using Lewis's theory), but that theory does not serve as a true description of the high-energy relation. For the high-energy case, we need a new theory that more fully describes this relation that approximates to the low-energy notion of causation that we are familiar with.

Lewis's theory is, on the current proposal, a bit like GR itself. It is an approximately true description of a certain energy regime, but ultimately not a fully general account of what we use it to describe—namely the relation that approximates the causal relation that we usually take to be part of the ontology of Lewis's theory. The question remains, on this view, as to what this causation-like relation is, and how it works. To answer this question, we may require a new theory, one that—like the various approaches to QG—provides an entirely new way of thinking about some aspect of the world. Put this way, QG not only promises to force a deep revision to the way that we think about the aspects of reality usually characterised by science but also suggests the need for us to revise how we think about the metaphysics of causation.

8.3.1.3. Which Option?

Ultimately, we don't think that there is a deep difference between the two responses just outlined to worries about high-energy causation. In both cases we deny that causation is emergent, and in both cases we accept that the notion of causation that is tracked by our current metaphysical theories must be replaced by a more fundamental notion, one that approximates the standard ways we think about things at low-energies. The central difference between the two responses is over whether we should call this more fundamental thing 'causation' or not. The first response has it that we should, the second response has it that we shouldn't. We are prepared to call this more fundamental relation causation, and so we adopt the first response, but not much hangs on this.

What matters is that causation *in a certain sense* does not exist: namely, the notion of causation that is the target of current metaphysics, and that might be loosely called a common-sense or everyday notion of cause. This is so regardless of whether there might be some more fundamental relation that we are willing to call causal. Note that even if we deny that causation in the everyday sense exists, we have argued that approximate spacetime has enough physical structure for us to be able to make sense of how empirical observation occurs; the physical structure at issue underwrites counterfactual dependence. So long as we can say, at the broadest level, that if some entity x had not occurred, then experiences of a certain kind would not have occurred, we have everything we need for the observation of entities. It is our contention that counterfactuals of this stripe are available and, indeed, that counterfactuals of this kind, within the low-energy regime where observation occurs, are true whether or not spacetime exists. Hence, we have all we need for observation and confirmation.

One might continue to worry, however, that the counterfactuals we usually take to support causation are, themselves, sufficient for the existence of a C-series. If so, then either the loss of spacetime does not imply the elimination of time, in all of the folk senses, after all, or, if it does, it also implies the elimination of true counterfactuals. But we know of no straightforward demonstration that the existence of a C-series is entailed by the presence of counterfactual dependence alone. This is mainly because counterfactual dependence is not transitive in the way that the causal relation—and indeed C-relations—need to be. So, while we accept that the counterfactuals at issue are true, and that their truth is indicative of a causation-like relation in the world (one that can support observation) we don't yet see why this should force a commitment to the existence of a C-series.

8.3.2. Generalising the Account

So far, everything we have said has been framed in terms of Lewis's particular counterfactual theory of causation. A similar picture of causation to the one that we have outlined here can potentially be developed for other accounts of causation. We admit, however, that this may be harder for those theories that are more tightly tied to spacetime. Consider, for instance, the conserved quantity or mark-transfer theory advocated by Salmon and Dowe. As previously discussed, according to the Salmon-Dowe picture, causation is the transfer of a conserved quantity between worldlines that intersect in spacetime; a picture of causation that is explicitly at odds with the notion that GR spacetime does not exist.

Still, most of what we need to say to support the conserved quantity picture can still be said even if spacetime does not exist. As before, GR remains an approximately true description of the low-energy regime at which the Salmon-Dowe theory is typically applied. Thus, we can describe the world as if the Salmon-Dowe theory were true, without getting ourselves into trouble. From the perspective of that theory, and with respect to the low-energy regime, there is no substantial difference in the way that the physics works, which guides that theory.

We encounter difficulties with the Salmon-Dowe picture when we try to extend it to a world without GR spacetime, given how closely tied to spatiotemporal concepts that picture ultimately is. But even here the theory is not beyond help. We can employ a very similar strategy to the one we used for Lewis's theory. We can replace talk of spacetime with talk of approximate spacetime. We then recognise that the conserved quantity theory is, at best, a characterisation of causation at low energies and for entities that can be approximately described in spatiotemporal terms. We then generalise the theory by allowing that, within the physical structure that ultimately corresponds to the ontology of some approach to QG, there may be paths or trajectories through the structure across which conserved quantities are exchanged, in some sense. Exactly how this works will require a good deal of development, and, will need close attention to the details of the physics. But we can see, in at least a rough-and-ready way, how the generalisation might go. We could also refuse to generalise the Salmon-Dowe theory and simply take it as a theory of causation to be replaced (or a causation-like relation that behaves as the theory says it should at low-energies), once again treating a metaphysical theory of causation in an analogous way to how GR may be treated (as an approximation).

8.4. Conclusion

We should take stock. In this chapter, we responded to a range of arguments in favour of the idea that a C-series is necessary for causation. We take ourselves to have undermined the most obvious reasons to suppose that this is so. Even if the C-series does not exist, it does not follow that there is no causation. We then proceeded to argue that the absence of the C-series is compatible with the existence of causation, at least in a certain sense.

We noted, however, that there is then a further question of whether causation should be considered emergent or not. We maintain that causation is not emergent: it is part of the fundamental structure of reality, though perhaps not in a form that is at all familiar to us. We recognise, of course, that the various accounts of causation that we have offered are likely to stretch the common-sense usage of the term, and maybe even our standard concepts of causation.

By recovering causation in the absence of a C-series we have shown that the loss of such a series presents no barrier to observation and empirical confirmation. Observation, we maintain, can be grounded in an entity that approximates GR spacetime. With the observation problem solved, we no longer have any reason to worry about the empirical coherence of an approach to QG that eliminates GR spacetime. We thus see little reason to suppose that GR space-time must be emergent, and thus that a C-series grounded in GR spacetime must exist.

Granted, it *could* turn out that the C-series arises from causal structure but there is no reason to suppose that this must be the case, and that is all we really need for the quick argument for timelessness to go through. We can therefore conclude in favour of our quick argument for timelessness. There are some approaches to QG that eliminate a C-series, and we don't yet have a reason to suppose that the C-series emerges within those theories. It remains a live epistemic possibility that the C-series does not exist. It follows from this that the discoveries that would lead people to reject the existence of time are not incompatible with our best science (modulo the qualification discussed at the start of Chapter Five and at the end of the last chapter, namely that people may not be responding to the loss of the C-series).

This concludes Part Two of the book. In Part Three, we return to the issue that animates our investigation of the folk concept of time, and that constitutes one of the three reasons to suppose that time must exist: agency. There, we build on the picture we developed in this chapter, according to which causation, in some good sense, exists even if GR spacetime does not. We show how causation, so understood, can scaffold agency. Thus, agency can exist without time.

PART III

9

An Error Theory About Time

9.1. Introduction

Here's the story so far. In Part One of the book we provided some much-needed clarification of the folk concept, or rather concepts, of time. We showed that there are certain discoveries that would lead people to reject the existence of time and so the everyday concept of time is not immune to error. In Part Two of the book we considered the idea that the kinds of discoveries that would lead people to reject the existence of time are incompatible with what we know about the world from science. We argued against this claim via the quick argument for timelessness, which seeks to establish the epistemic possibility of a world without a C-series. Our defence of the quick argument for timelessness ended up drawing heavily on causation. It is epistemically possible to discover that the world lacks a C-series, but only if there can be causation without time.

In the remainder of the book we consider one final reason to suppose that time must exist. The idea, in a nutshell, is that the discoveries that would lead people to reject the existence of time are incompatible with the existence of agency. Specifically, without at least a C-series, there cannot be any agency. Given that we know that agency exists, it follows that discovering there is no C-series is at odds with what we know. It is thus unthinkable that we could make the sorts of discoveries that would in fact lead people to deny that time exists, and so the absence of time is not a viable possibility.

Our goal is to argue that the discoveries that would lead people to reject the existence of time are, in fact compatible with agency. Before we get into the details of the argument, however, it is important to outline some assumptions that we're making. First, as with Part Two of the book, we focus on the presence or absence of the C-series, since it is the absence of the C-series that we believe to be responsible for people's judgements about the non-existence of time. This focus on the C-series is subject to the caveat that we discussed in Part One. Namely, that there could be some other feature of the vignettes used in our studies that lead people to deny that time exists other than the absence of the C-series. Since we don't know what this other feature

Out of Time: A Philosophical Study of Timelessness. Sam Baron, Kristie Miller, and Jonathan Tallant, Oxford University Press. © Sam Baron, Kristie Miller, and Jonathan Tallant 2022. DOI: 10.1093/oso/9780192864888.003.0009

might be, however, we have no idea whether losing that feature is compatible with agency. As with a similar issue in Part Two, this plays in our favour. For it is hard to see how one could make the argument that agency can't exist in the absence of whatever it is that leads people to reject the existence of time, if we don't know what that feature is. Any reason based on agency to show that time must exist, would thus seem to be highly questionable.

The second assumption we make is that the issues to do with agency arise downstream of the issues to do with science discussed in the previous chapter. We thus carry forward some of the results from Part Two into Part Three. In particular, we assume that there can be scientific discoveries that lead people to deny that time exists only if there can be causation without time (for if there cannot, then it seems that spacetime must be emergent on pain of the empirical incoherence of the theories that promise to yield such discoveries).

Third, we will adopt the modest proposal for causation without time outlined in Chapter Eight, and thus assume that accompanying any discovery that time does not exist is the discovery that something approximating space-time exists, and thus that the existence of approximate spacetime is not sufficient for time to exist. We also set aside the idea that approximate space-time might be used to recover something that satisfies the various folk notions of time. Perhaps it can (as discussed in Chapter Seven), but it is epistemically open that approximate spacetime exists without time. We set this aside because if approximate spacetime yields time, then there is little reason to consider any argument from agency against the idea that time might not exist. For obvious reasons, the discussion in Part Three would be otiose.

With these assumptions in mind, we will now turn toward agency. Rather than tackle agency directly, we will work up to it via a discussion of temporal thought and talk. We do this because it is plausible that the relationship between agency and folk conceptions of time is mediated by temporal thought and talk. After all, it is natural to suppose that if our folk concepts of time are in error, then our temporal discourse is error theoretic, and, in turn, that any discourse that is founded on an error theoretic temporal discourse, or which presupposes the truth of that discourse, is also error theoretic. Since it is *prima facie* plausible that our agentive discourse does indeed presuppose that our temporal discourse is in good standing, this raises the startling prospect that our agentive discourse will be in error, if our temporal discourse is in error. In order to consider agency, then, we need to start by looking at temporal thought and talk. That is what we do in this chapter and the next. In Chapter Eleven, we return to consider agency. There we argue that agency could survive the discovery that there is no time, and thus survive the discovery that our temporal thought and talk is in error, despite appearances to the contrary.

Our first goal in this chapter is to spell out the connection between the discovery that our folk concepts of time are in error and the status of our temporal discourse. Is it in fact the case that if our folk concepts of time are unsatisfied, then our temporal discourse is error theoretic? In order to answer this question, we need to take a step back and explicate temporal error theory.

In §9.2 we ask, 'what is temporal error theory?' There, we situate the discussion of error theory within a broader framework of options. In §9.3 we then argue that in order for our temporal thought and talk to be in good standing while our folk concepts of time remain unsatisfied, one must find a plausible way to sever the connection between the satisfaction of the folk concepts of time, and the truth of our temporal thought and talk. In §9.4 we consider temporal non-cognitivism, which severs this connection by maintaining that our temporal thought and talk is not truth-apt. We argue that the prospects for developing a viable version of temporal non-cognitivism are poor. In §9.5 we then consider two versions of temporal realism. These views both attempt to sever the connection between the satisfaction of our folk concepts of time, and the truth of our temporal thought and talk, though they sever that connection in somewhat different places. By the end of this chapter we will have argued that if our folk concepts of time are not satisfied, then temporal error theory is most likely true.

9.2. Temporal Discourse: The Options

We have plenty of temporal thoughts. We think that the meeting will begin in 5 minutes, that the drive will take 3 hours, that Sara's birthday is on 31 July, that the dinosaurs were extinct by 300 BC, that the extinction of the dinosaurs was earlier than the extinction of the New Zealand Moa, and so on. Let's call these thoughts, collectively, our 'temporal thoughts'. Jointly, the utterances we use to express them may be dubbed our 'temporal discourse'. These thoughts, (and hence the utterance we use to express them) appear to be endangered by the discovery that our folk concepts of time are unsatisfied. After all, our temporal thoughts employ temporal concepts, including the concept of a temporal relation (earlier-than and later-than or betweenness), the concept of a location in time (past and future), the concept of a duration of time (5 minutes) and so on. Temporal thoughts, one might say, are *about* time, because they employ temporal concepts. So, if our folk concepts are in error, then our temporal thoughts are in error (likewise for our temporal discourse).

This brief argument, however, is too swift. For there are, in fact, three potential responses to the discovery that our folk concepts of time are in error:[1]

> **Temporal Realism:** ordinary temporal thought is truth-apt, and many core thoughts in that domain are true.
>
> **Temporal Error Theory:** ordinary temporal thought is truth-apt and core thoughts in that domain are not true.
>
> **Temporal Non-Cognitivism:** core thoughts in the temporal domain are not truth-apt because such thoughts are not cognitive.

Temporal error theory is analogous to moral error theory. Very generally, to be an error theorist about some domain of discourse, D, is to hold that the domain of discourse is both truth-apt and false. We're prepared to go along with Olson and treat standard versions of moral error theory as the view that:

> ... moral judgements are assertions that attribute mind-independent (but non-instantiated) moral properties to objects and that, as a consequence, moral judgements are systematically mistaken   (Olson 2014: 4).

Or, to be a little more 'to the point' about matters, 'all positive moral judgements are false (or at least not true)' (Sinclair 2012: 158).

If *moral* error theory is the view that moral judgements are assertions and hence are capable of being true or false, but are systematically false (or, at least, *not true*), then *temporal* error theory ought to be the view that *temporal* judgements are assertions that are capable of being true or false, but are systematically false (or, at least, *not true*). At a first pass, at least, that looks a natural way in which to generalise the error-theoretic insight.[2]

However, because our focus is on temporal concepts, we are not entirely happy with the focus on temporal *talk* as a way of developing temporal error

---

[1] Within error theory there are two further positions: fictionalism and eliminativism. In the end, then, we are left with four options with regard to temporal thought: temporal fictionalism, temporal eliminativism, temporal non-cognitivism, and temporal realism. We introduce and discuss fictionalism and eliminativism in the next chapter.

[2] And, of course, the consequences of holding such a view are going to be quite broad. It would seem natural, after all, to suppose that talk that is implicitly or explicitly about any of past, present and future would all be, in a sense, *temporal*. If it then turns out that a temporal error theory is correct then just about all of our discourse will turn out to be truth-apt, but false.

theory. Thus, we focus on temporal error theory as a claim about *thought* rather than *talk*. So, just as the domain of moral thought is one about which one can be an error-theorist, so too is the domain of temporal thought.[3] We assume that if temporal error theory is true of temporal thoughts, then it is also true of temporal talk.[4] *Prima facie,* temporal thoughts employ temporal concepts. As such those thoughts are about time.

The temporal realist agrees with the temporal error theorist that our everyday temporal thought and talk is truth-apt. However, our everyday thought and talk about time is, she says, true. The temporal non-cognitivist, by contrast, disagrees with both the realist and the error-theorist. She maintains that our everyday temporal thought is not even truth-apt, and so cannot be either true or false.

9.3. From Concepts to Error Theory?

What is the link between folk concepts of time and temporal error theory? We saw in the first part of this book that there is no single folk concept of time. So, different individual's temporal thoughts will employ different folk concepts of time. Given this, it is not quite right to simply talk about temporal error theory. Rather, it seems more likely that 'temporal error theory' might be a family resemblance term for a number of related and overlapping error-theoretic accounts that make use of a number of related and overlapping folk concepts of time. Nonetheless, as we also saw, there are discoveries that could be made about the world that would lead a majority of people studied to deny the existence of time, regardless of the more specific concept they have.

---

[3] A domain of thought, roughly speaking, is error theoretic just in case the thoughts in that domain are truth apt, and the thoughts in the domain are false, or at least, not true. In the background to this discussion are some complicated issues around the content of thoughts, judgements, assertions and sentences. Our assumption in what follows is that sentences, thoughts, and other expressive devices all express or denote some contents. We'll call these contents *propositions*. We don't make any claims as to the nature of these things. We will also sometimes elide the distinction between what is said or written, and what is thereby expressed. Nothing turns on any of this.

[4] One might worry that in a world that lacked time there could be no temporal thoughts (let alone temporal discourse): for one might think that thoughts are themselves temporally extended, or that the having of contentful thoughts, even if those are not temporally extended, requires the existence of time. So you might think that the claim that our world is genuinely timeless is undermined by the fact that you are currently having some contentful thought. Consequently, you might think that the question of whether or not our world is timeless (and what it would take to be timeless) is uninteresting, since a perfectly good transcendental argument shows is that it is not timeless. Again, we think these are pressing concerns, and something to which we return in later chapters. In the meantime, we want to just accept this as a broad characterisation of temporal error theory, and move on to consider our folk concept of time. After all, even if our world is not timeless, and is obviously not timeless, the question of what it would take for a world to be timeless would still be an interesting one.

Shared temporal error theory is the idea that the majority of people's thoughts about time are in error, because some such discovery has been made.

Here is one way to understand the relationship between shared temporal error theory and our folk concepts of time. Temporal thought, in general, is true only if the temporal concepts that are employed in temporal thought are satisfied. The natural thought here is that these temporal thoughts are about time, and so they are only true if time exists. Then, on the assumption that most of our folk concepts of time are in error, shared temporal error theory will be vindicated. According to this line of reasoning, a shared folk temporal error theory is true when temporal thought is systematically false for a population regardless of which concept of time individuals in that population employ. (Henceforth, we will simply talk of shared temporal error theory, omitting talk of that error theory being an error theory about folk thought and talk).

The proposed connection between the folk concepts of time and shared temporal error theory is quite strong. Whether the connection is plausible depends on two further facts. First, is everyday thought and talk about time *truth-apt*? If it is not, then the connection between the folk concepts of time and temporal error theory is severed. The fact that the folk concepts of time are not satisfied cannot show that our everyday thought and talk about time is false, because it isn't even truth-apt.

Second, if our temporal thought and talk *is* truth-apt, can we resist the idea that it is *about* time in one or all of the folk senses? If we can, then perhaps realism about temporal thought or talk should be retained even if our folk concepts of time are shown to be in error. That is, is there a way to insulate the truth of our temporal thought and talk from the fact that nothing satisfies folk conceptions of time?

In what follows we take up these questions. We do this by considering both folk temporal non-cognitivism and folk temporal realism in more detail. We argue against both views, and thus in favour of the strong connection, proposed above, between the folk concepts of time and folk temporal error theory.

9.4. Temporal Non-Cognitivism

As noted, temporal non-cognitivism is the view according to which our temporal thought and talk does not involve truth-apt beliefs, but rather, involves some other kind of attitude. So, even if our folk concepts of time are not satisfied, it does not follow that such talk is false because *a fortiori* it is neither true nor false.

We find temporal non-cognitivism tantalising. We confess, though, that we are not sure how to spell out the view in a plausible manner. Given this, what we say about temporal non-cognitivism will be brief. Our aim is not to argue that there is no way to develop a version of temporal non-cognitivism. Rather, we will gesture towards what we take to be some powerful problems that the non-cognitivist must face if she is to develop such a view.

The central problem we see for temporal non-cognitivism is that it is difficult to see what our temporal thoughts might be, if not beliefs,[5] and hence which mental states temporal talk might express, if they are not reporting beliefs. Indeed, we can only really see one way to develop this idea, and that is to take temporal thought and talk to express non-cognitive attitudes of some kind. Call such a view: temporal expressivism. The question then becomes *which* non-cognitive states the expressivist might identify with our temporal thoughts.

There are roughly two kinds of attitudes that one might appeal to: emotive attitudes such as regret, anticipation, nostalgia, fear, anxiety; or evaluative attitudes such as desire or preference. These are natural options to pursue, because they all display a 'temporal' asymmetry (Callender and Suhler 2012) and so bear some connection to time already.

Any plausible temporal non-cognitivism will need to provide felicity conditions for temporal thoughts such as 'it rained yesterday' and 'it rained 5 minutes ago'. After all, even the non-cognitivist about temporal thought presumably doesn't want to say that any temporal thought is as apt as any other. We don't want Freddie's thought that it rained 5 minutes ago in Hobart, and your thought that it did not rain 5 minutes ago in Hobart, to be equally apt, even though according to the non-cognitivist neither thought is true (because not truth-apt). It's not clear, however, how to go about providing felicity conditions for temporal thoughts. If Freddie has a 5-minute attitude towards an instant in which it rains in Hobart, and you have that same attitude towards an instant in which it does not rain in Hobart, in what sense, if any, is only one of the thoughts apt?

So, the non-cognitivist will want to say something about the conditions under which the relevant non-cognitive attitudes are apt, and the conditions

---

[5] Deng (2018) links the temporal and normative cases together quite closely. The account of temporal ontology that Deng offers may provide the resources needed to develop a version of non-cognitivism.

under which they are not apt. Now, it would be natural to say that what makes these sorts of attitudes apt, or not apt, is whatever it is that satisfies our folk concept(s) of time. Whatever that thing is, it rationalises our having certain attitudes and not others. But that, of course, is not something the non-cognitivist can say in this case, given that we are supposing our folk concept(s) of time are in error.

One potential way to address this issue is to appeal to the notion of approximate spacetime. Thus, while there is no ordinary temporal structure that can provide felicity conditions for the attitudes that underwrite non-cognitivism, perhaps there is enough structure in approximate spacetime to underwrite the relevant attitudes. On this view, then, the attitudes would be 'time-like' attitudes, in so far as they are responsive to approximate spatio-temporal structure.

One concern we have for this option relates to the concept of time. Suppose that the folk deny that time exists because there's no C-series. We are assuming that approximate spacetime is not sufficient for satisfying the folk concept of time. After all, if approximate spacetime gives us enough to satisfy the folk notion of time, then there's little need for expressivism in the first place. We can just revert to an ordinary variety of realism about temporal thought and talk.

The expressivist we are imagining, then, is someone who takes our folk notions of time to be in error, but who nonetheless deems attitudes like regret and anticipation to be felicitous (because of some other underlying structure). It seems plausible, however, that if one were to take these attitudes to be felicitous, then one would also take one's concept of time to be satisfied and thus not in error. It is difficult to square the view that our temporal attitudes are generally felicitous with the idea that the folk notion of time is not satisfied.

Even if felicity conditions for some temporal attitudes can be provided along these lines, there is a deeper problem for temporal expressivism. The problem is a paucity of attitudes. Whichever attitudes one chooses—emotive or evaluative—there do not appear to be sufficiently many, or sufficiently fine-grained, non-cognitive attitudes to capture all of the different temporal thoughts we have.[6] To illustrate: we can think 'it was raining an hour ago',

---

[6] One might think that there 'evidential attitudes', like degree of confidence or acceptability (see Jaszczolt 2013). However, we take it to be relatively clear that such attitudes are cognitive. A degree of confidence is pretty plausibly a degree of belief, which has an essential doxastic component. A degree of acceptability might be non-cognitive, depending on what 'acceptability' is, but if 'acceptability' is an evidential notion, then it likely means: degree of evidential support, or something like that. But this also appears doxastic: it is the degree to which one believes the evidence supports a claim.

'it was raining two hours ago', 'it was raining three hours ago', and so on. Since the object of these thoughts is the same in each case—that it rains—what distinguishes them must be the conative attitude we have towards that state of affairs. But there aren't sufficient attitudes to allow us to distinguish the three thoughts listed, let alone all of the thoughts we can have about its having been raining at past times. In general terms, our temporal thought and talk is much richer, and much more detailed than the structure of attitudes.

Of course, this issue can be managed to some extent by allowing that attitudes come in degrees. So, for instance, one can say that there are degrees of anticipation, degrees of regret and so on. Even so, there appears to be a mismatch between how fine-grained attitudes, like anticipation can be, and how fine-grained temporal thought can be. In principle, one can have temporal thoughts that are extremely fine-grained, down to the scale of seconds, or nanoseconds. It is implausible, however, that we can place each such fine-grained temporal thought into a one-to-one correspondence with some similarly fine-grained attitude. The main problem is that we can't differentiate degrees of anticipation to the same extent. Can one really have a greater degree of anticipation for an event in a second, as opposed to one in two seconds? What about nanoseconds? It seems doubtful. Attitudes are just not that finely structured.

We don't say that it is not possible to provide a satisfactory non-cognitivist account, only that it's not obvious how to do so. It's particularly difficult to evaluate the prospects here, given that we don't have anything like a full account of the attitudes that are relevant in developing temporal non-cognitivism.[7] For now, at least, the non-cognitivist route seems to us to be a difficult way forward.

In what follows we will therefore assume that temporal thought is truth apt, and so is either true, or is false. Let's first consider the option according to which it is true.

9.5. Temporal Realism

In the face of the discovery that nothing satisfies the folk concepts of time one might still think that temporal realism is much more plausible than temporal

---

[7] The temporal noncognitivist will also need to solve the Frege-Geach problem: the problem of attributing truth-values to complex thoughts containing both cognitive and non-cognitive components. Perhaps the temporal noncognitivist can, at this point, borrow from her moral non-cognitivist counterpart. After all, this is a perfectly general problem facing noncognitivists. We certainly do not want to foreclose this possibly. We do think, however, that in order to do so she will probably need to at least provide felicity conditions for temporal thoughts.

error theory. A natural thought is that even though our world's structure does not satisfy our folk concepts of time, nevertheless that structure is somehow sufficient to vindicate our temporal thought and talk. That would be to suggest that our temporal thought and talk has a different conceptual foundation than we otherwise supposed: it is made true by features of what exists, even though what exists is not temporal by the lights of any folk conception.

In the face of our folk concepts of time, collectively, being unsatisfied, the realist about temporal thought and talk would need to sever the connection between the satisfaction of these folk concepts of time, and the truth of our temporal thought and talk. As we see it, there are two options available for severing this connection: non-aboutness realism and hermeneutic realism.

Non-aboutness realism is the view that our temporal thoughts are made true by something *other* than what they are about. To see how this might go, consider the mereological nihilist for a moment. The mereological nihilist holds that there are no composite objects.[8] What, then, of our composite-object thought and talk: our talk about tables, and chairs, and so on? The nihilist might concede that our table thoughts, say, are indeed about tables, and that tables, if there were any, would be composite objects. So, our table thoughts are about something—tables—that do not exist. But she might argue, our table thoughts are nonetheless made true, and they are made true by something other than tables: namely, by simples arranged table-wise. To our knowledge, no one has ever developed quite this view about table-thought.

Nonetheless, it's quite notable that a view like this is common in the philosophy of time. Presentists, for instance, hold that there are no past things. They also hold that we have thoughts such as 'there were dinosaurs'. Most presentists concede that these past-tensed thoughts are about past things: in this case past dinosaurs. But there are no past dinosaurs. Nevertheless, many presentists argue that these thoughts are true despite the fact that the things they are about do not exist: that is because they are made true by some presently existing proxy such as a thisness, or a primitively tensed property of the world, or some such.[9]

We have already seen, in Chapter Eight, that viable versions of QG that eliminate space-time will be ones that countenance a structure that approximates spacetime at certain energy levels. Given this, the non-aboutness realist could argue that although our temporal thoughts are not about the

---

[8] Defenders of mereological nihilism include Brenner (2015), Contessa (2014), Dorr (2002), Liggins (2008), and Sider (2013).

[9] Ingram (2016, 2019), Tallant and Ingram (forthcoming), Bigelow (1996), Keller (2004), Crisp (2007), Cameron (2011).

approximating structure, they are nevertheless made true by that structure. There is, of course, a good deal of work to be done in spelling out the truth-conditions for our temporal talk in terms of facts about something that approximates spacetime. Still, we hope that the reader can see, at least in principle, how we might begin such a process. The idea would be that whatever truth-conditions we would have proffered for our temporal thought and talk, in terms of times and temporal relations, can simply be swapped out for truth-conditions involving relevant parts of the approximating structure.

Note that this view is *not* to be confused with the view that our temporal thought is about approximate spacetime, and approximate spacetime makes that thought and talk true. Given the proposed link between temporal thought and the folk concept of time, such a view would need to be one on which approximate spacetime is sufficient for the folk concept of time to be satisfied. After all, our temporal thought is about whatever satisfies the folk concept, and so if approximate spacetime does that work, then that's what our temporal thought is about.

As noted at the beginning of this chapter, the view that approximate spacetime is sufficient for the concept of time is one that we are setting aside for the time being, based on our discussion in Part Two. Given this assumption, the only way to get temporal thought and talk to be made true by approximate spacetime given that it is not sufficient for the folk concept of time to be satisfied (and assuming that the folk concept is not satisfied) is to sever the link between what temporal thought is about and the folk conceptions of time. The only viable form of realism, then, is one according to which our thought and talk is about whatever does satisfy the folk concept (such as a C-series) which is something not provided by approximate spacetime, and our thought and talk is nonetheless made true by approximate spacetime structure. We think it is important to differentiate the non-aboutness view from the view that approximate spacetime provides structure enough to satisfy the folk concept of time, since we suspect that the initial plausibility of the former view is based on mistaking it with the latter view.

As noted, the alternative to non-aboutness realism is revisionary hermeneutic realism. This view is a kind of *hermeneutic* realism because it's a view according to which the best interpretation of our temporal thoughts is one on which those thoughts are not about time, or temporal relations, at all. The view is *revisionary*, in that (we assume) the defender of such a view allows that our temporal thoughts *appear* to be about time, and temporal relations. So, the view is asking us to change our beliefs about what our temporal thoughts are *really* about: relevant parts of the approximating structure. Those thoughts are made true by those aspects of that structure. The revisionary hermeneutic realist, then, can appeal to the very same truth-conditions proffered by the non-aboutness realist (and *vice versa*).

Both versions of realism are attractive insofar as they vindicate our temporal thoughts. Both will agree about which of our temporal thoughts are true, and which false. The only real difference between the views is that the non-aboutness realist says that our temporal thoughts are made true by something other than what they are about, while the revisionary hermeneutic realist says that they are made true by what they are, in fact, about—they're just not about what we originally thought. Nevertheless, there are clear costs to each view. Since these seem to be the only realist options available, we don't really see much hope for temporal realism. Consider, first, non-aboutness realism.

First, we think it unlikely that non-aboutness realism will be empirically vindicated by the kind of evidence offered in Part One of the book. What this view requires is that even though people will respond that there is no time in Condition C, nevertheless in that self-same condition they will respond that their various temporal thoughts and talk are true. To be clear, we don't have empirical evidence on this matter. But we are deeply sceptical that people would in fact respond in the relevant way. Indeed, it seems to us that there is a sort of catch-22 faced by this view, and some others like it.

To see this, return to mereological nihilism for a moment. Suppose that people are indeed inclined to say that their table talk is true, even though there are only simples arranged table-wise, and no composite objects that are tables. The very people who judge that table talk is true in that eventuality are, we suspect, people who will judge that tables just are simples arranged table-wise. Conversely, we find it plausible that those who really do judge that there are no tables if there are only simples arranged table-wise will judge that their table talk is, in that eventuality, not true. The reason for this is that a table-wise arrangement of simples is a reasonable candidate to satisfy our concept of a table. So, it seems very likely that people who judge that their table talk is true, will thereby judge that this is what tables turn out to be.

Insofar as one thinks that whatever approximates spacetime makes true our temporal thought and talk, it seems to us, one is very likely to think that such a structure satisfies one's folk concept of time. In that case, then, temporal realism accompanies realism about time itself. Conversely, insofar as one has a folk concept of time that is not satisfied by that structure (which, it will be recalled, is the situation we are considering, because the structure at issue does not support a C-series), then one will be inclined to say that one's temporal thought and talk is false.

Suppose, however, you don't buy this reasoning: perhaps you think our empirical speculation is false. Still, there are costs to the view just in virtue of its commitment to non-aboutness. Non-aboutness realism dictates that our temporal thoughts are made true by something other than what they are about. Many find this kind of non-aboutness approach unappealing, whatever its domain, because it potentially leaves many truths unmoored from reality.[10] The worry, in brief, is that truthmaking is constrained by aboutness. What makes a claim true ought to be just what that claim is about. Non-aboutness realism requires giving up this plausible constraint on truthmaking. Once this is given up, however, it is hard to resist the idea that just about anything can count as a truthmaker for any proposition. Aboutness plays a crucial role in constraining the kinds of things that can count as a truthmaker for a given claim (see Merricks (2007) for an extended defence of this idea). Presentism, for instance, comes in for a drubbing for holding that past-tensed truths are made true by something other than that which they are about. There are responses to these sorts of worries,[11] but it is unclear that such responses manage to fully mitigate the cost of the view. This gives us reason to try to find an alternative to non-aboutness realism.

What, then, of revisionary hermeneutic realism? On the one hand this view has a general advantage over non-aboutness realism; it allows us to say that what makes our thoughts true is what those thoughts are about. Still, although on this view what makes our temporal thoughts true is what they are about, what they are about is not what they *appear* to be about. According to revisionary hermeneutic realism, it turns out that our temporal thoughts are about something quite different from what we took them to be about, and that only after appropriate reflection and interrogation of our practices can we be brought to see that our temporal thoughts are about a non-temporal structure. So, the view is committed to what some might see as the counterintuitive consequence that at least sometimes we don't know what our thoughts are about.

The revisionary hermeneutic realist appears to be committed to the view that sometimes we don't know the internal natures of the things our thoughts are about.[12] For she thinks that we don't even know whether or not our

---

[10] See for instance Caplan and Sanson (2010), Merricks (2007), Baron (2013), Cameron (2008).

[11] For discussion of these issues see Tallant and Ingram (2020: Section 2); Crane (2013), Baron, Chua, Miller and Norton (2019).

[12] If you're an externalist, then you think that what our thoughts are about is partly determined by which things in the world each of us is in causal contact with. If so, then you naturally think that

temporal thoughts are about time, not just that we don't know the internal nature of the thing our temporal thoughts are about.

Now, there might be a response available on the part of the revisionary hermeneutic realist. She might argue that we know that our temporal thoughts are about *that* phenomenon, where we point (metaphorically) to whatever phenomenon is, say, responsible for the appearance of temporal duration, temporal order, past, future, and so on. We don't know what the internal nature of the thing is, that lies behind those appearances, and we just assume that whatever we are pointing to is time. In fact, though, it turns out not to be time after all; but instead to be some other structure. Still, we know what our temporal thoughts are about in some good sense: they are about *that* phenomenon, it is just that we don't know much about that phenomenon, which is compatible with it turning out to be the case that the phenomenon at issue is not temporal.

We know, from our empirical studies, that there are ways that the phenomenon—the one responsible for the appearance of temporal duration, order, and so forth—can turn out to be, such that people judge that there is no time. So far, that is consistent with revisionary hermeneutic realism. After all, the revisionary hermeneutic realist holds that our temporal thoughts are true, but not in virtue of temporal structure; so, she should allow that there are conditions under which we would judge that there is no time, and in which we would nonetheless judge that our temporal thoughts are true.

Note that hermeneutic realism is not to be confused with the idea that temporal concepts are somehow altered so that they come to be satisfied so long as certain appearances obtain. Hermeneutic realism is a claim about our temporal thoughts, and the idea is that such thoughts are shown to be disconnected from our temporal concepts in such a manner that the relevant thoughts can be true even though the folk concept of time is not satisfied by anything, and people are thus inclined to say that time does not exist.

As we see it, there are two difficulties with hermeneutic realism. First, putatively temporal thoughts are, by virtue of the revision, disconnected from our everyday temporal concepts. It is just not clear, then, in virtue of what the relevant thoughts can be genuinely called temporal thoughts. Perhaps

---

sometimes we don't know the internal natures of the things about which we are thinking. In this sense, then, we don't know what our thoughts are about. Our water thoughts, for instance, weren't known to be about H_2O, even though that is what they were about, before we knew that water is H_2O. Still, we knew *something*: we knew that our water thoughts were about that stuff over there (insert an appropriate demonstrative here) even if we didn't know the internal nature of that stuff. We knew they were about *water*.

they have the surface structure of temporal thoughts, insofar as they can be modelled by propositions that involve temporal relations, but in a deep sense the thoughts are not about time, because they don't invoke any temporal concepts. The first worry for the hermeneutic realist, then, is that they've simply changed the subject: the relevant thoughts are, post-revision, no longer temporal in a recognisable sense.

This leads to the second difficulty. Ultimately, our interest in temporal thoughts stems from the putative role that they play in scaffolding agency. In general, it is difficult to see how the hermeneutic realist's revised thoughts can continue to play this role post-revision. For the relevant thoughts don't seem to be about time any longer, and it is thoughts about time that appear, on the face of it, to be implicated in agency. As we shall also see, there appears to be a tight conceptual connection between the concept of time and the concept of agency. Since revisionary hermeneutic realism does nothing to address this conceptual connection, leaving it in place, it cannot fully address the problems we articulate for agency in Chapter Eleven.

We thus don't see non-aboutness realism or hermeneutic revisionary realism as viable approaches to temporal thought. We should note, however, that we are happy to be wrong about this. For recall that, on both views, our temporal concepts are assumed to be in error. Given that, if our concepts of time are in error, then people would deny that time exists and either realist approach would be *compatible* with the view that there is no time. If these realist approaches could get us agency, then we would have a rather swift way to reach the desired conclusion of Part Three of this book: namely, that there can be agency without time. In Chapter Eleven, we offer a different approach that reaches the same goal. So long as there is some approach that works, however, that's all that really matters.

Once again it is worth emphasising that the realist views discussed here are not to be confused with nearby views, on which the folk concept of time is satisfied by, say, approximate spacetime. It is also worth noting that even if approximate spacetime does satisfy the concept in this way (we are doubtful, see the discussion in Chapter Seven), there might still be need for the realist approaches sketched here. For temporal thought does seem to explicitly token temporal relations of the kind found in a C-series, and that does seem to be missing even when approximate spacetime exists. So, we may still need to try and reassess the felicity of temporal thought in this situation. In that case, we would recommend revisionary hermeneutic realism due to lingering worries concerning aboutness and truthmaking.

One final point. It is conceivable that, upon making the kinds of discoveries that would lead people to deny that time exists, people will in fact just adopt something like non-aboutness realism, or revisionary hermeneutic realism. This is a different empirical bet to the bets considered in Chapter Two. There we considered the possibility that people might continue to say that time exists upon making certain discoveries, despite the fact they would report doing something quite different. This kind of bet is a bet about the concept. The kind of bet we are considering here, is one about temporal thought. There are thus really two situations to consider. If the idea is that people would adopt non-aboutness realism or revisionary hermeneutic realism *even though* they deny that time exists, then that's fine. It would, as noted, be a way for us to get what we want. If, however, people accept that their concept of time is satisfied and they accept anti-aboutness realism or revisionary hermeneutic realism, then that's a problem for us, but the problem arises much earlier than our discussion of temporal thought. The problem arises back in Part One, for this would undermine our argument against the claim that the folk concept of time is immune to error. Either way, the new bet about temporal thought poses no distinct problem for us.

9.6. Conclusion

So far, we have argued that neither non-cognitivism nor realism are attractive accounts of our everyday temporal thought and talk. This motivates our search for an alternative. That alternative is temporal error theory. In Chapter Ten we investigate temporal error theory in its two guises: temporal eliminativism and temporal fictionalism. We argue that temporal eliminativism, too, is unappealing. Of the two kinds of error theory, then, this gives us reason to endorse temporal fictionalism.

We then consider a worry faced by both versions of temporal error theory: namely that if temporal error theory is true, then none of our agentive thoughts are true. But if none of our agentive thoughts are true, then we cannot choose to eliminate (as eliminativism requires) or choose to pretend (as fictionalism requires). So, temporal error theory is self-undermining. If that worry cannot be resolved, this suggests that even though realism, in two guises, faces certain difficulties, it nevertheless might still be the best of a bad lot. We resist this thought, and instead in Chapter Eleven present a solution to the worries facing error theory. There, we defend the combination of temporal fictionalism with realism about agentive thought.

10

The Trouble with Error Theory

10.1. Introduction

In Chapter Nine we outlined a range of responses to the discovery that nothing satisfies the folk concepts of time. Broadly speaking, there are three options: temporal realism, temporal error theory and temporal non-cognitivism. So far, we have argued that both non-cognitivism and realism face problems. That is not to say, of course, that temporal error theory faces no problems of its own.

Our goal in this chapter and the next is to outline a viable version of temporal error theory. In this chapter, we differentiate between two versions of temporal error theory: fictionalism and eliminativism and then raise a problem for both views. The problem is connected to the potential that error theory has to falsify agentive thought but runs deeper. Not only is agentive thought potentially falsified by temporal error theory, but this falsification threatens to make both forms of temporal error theory pragmatically incoherent in an interesting sense: one can't actually act in the ways that error theory seems to demand.

Throughout this chapter we generally assume a strong connection between temporal thought and agency. This connection is such that if temporal thoughts are false, then agentive thoughts are false as well. For now, we take our agentive thoughts to just be our thoughts about what we should do morally or prudentially, and what we can do to bring about certain desired ends. They are thoughts generally of the form 'in order to achieve x I must do y' (we will have more to say about these thoughts later on). As we will see, this connection is the source of the problems for both forms of error theory. In the next and final chapter, we weaken this connection in a way that allows us to recover agency, and then settle in favour of fictionalism about temporal thought.

We begin by briefly distinguishing temporal fictionalism and temporal eliminativism (10.2). We go on to consider each approach in turn, starting with eliminativism (10.3) before moving on to fictionalism (10.4). We conclude by diagnosing a common problem that both views face, in virtue of their error theoretic foundations (10.5).

Out of Time: A Philosophical Study of Timelessness. Sam Baron, Kristie Miller, and Jonathan Tallant, Oxford University Press.
© Sam Baron, Kristie Miller, and Jonathan Tallant 2022. DOI: 10.1093/oso/9780192864888.003.0010

10.2. Fictionalism and Eliminativism

As discussed in Chapter Nine, temporal error theory can be summarised as follows:

Temporal Error Theory: ordinary temporal thought is truth-apt and core thoughts in that domain are not true.

Temporal fictionalism and temporal eliminativism agree on the core error-theoretic claim that temporal thought and talk is truth apt and false. What they add to error theory is a normative recommendation: the eliminativist maintains that we should eliminate temporal thought and talk; the fictionalist maintains that we should retain temporal thought and talk and engage with it *as a fiction*. Thus, we have:

Temporal Eliminativism: (i) ordinary temporal thought is truth-apt; (ii) core thoughts in the temporal domain are not true, and (iii) ordinary temporal thought ought to be eliminated.

Temporal Fictionalism: (i) ordinary temporal thought is truth-apt; (ii) core thoughts in the temporal domain are not true, but nevertheless, (iii) ordinary temporal thought constitutes a useful fiction, and one that we should continue to engage with.

The distinction between fictionalism and eliminativism is familiar from the moral case. Moral eliminativism is the view that we should eliminate moral thought and talk; moral fictionalism[1] is the view that we should continue to engage in moral thought and talk by treating it as a useful fiction. In the moral case, arguments in favour of fictionalism are standardly developed as follows. First, one identifies a shortcoming with eliminativism: a cost associated with eliminating the domain of thought and talk at issue. The cost is then converted into a benefit of fictionalism, and thus used as justification for adopting a fictionalist attitude. In what follows we will try to argue against temporal eliminativism in a similar way. Ultimately, however, we will show that both temporal fictionalism and temporal eliminativism face a common problem.

---

[1] For a discussion of moral fictionalism see Joyce (2011), Chrisman (2007), Kalderon (2005), and Nolan, Restall and West (2005).

10.3. Temporal Eliminativism

Eliminativism about a domain of thought is the view that we should eliminate the thoughts within that domain. The reasoning that proceeds from error theory to eliminativism is often compelling. For example, we once held various thoughts about witches (and indulged in a discourse to accompany those thoughts). Many women were burned or drowned. But there were no witches. When we discovered that there were no witches, we discovered that our witch thoughts were systematically in error, and as a result, we concluded that we should eliminate those thoughts and abandon witch discourse.

More generally, we'd expect that if we discover some domain of thought to be completely in error, then we should eliminate those thoughts that lie within the relevant domain. After all, in *general* it's not a great idea to have false beliefs, and, on that basis, to engage in a systematically false discourse. We do better when our beliefs are true, because they appropriately track the way the world is. Moreover, and to return to the example, it's very clear that witch discourse was not a good thing: we don't want to burn and drown people, so we have excellent reason, both moral and prudential, to eliminate that discourse and those thoughts.

In theory, eliminativism about temporal thought is straightforward: it is the view that we should eliminate our temporal thoughts. In practice, though, it's difficult to get a handle on just how radical such a view really is. Temporal thoughts pervade so much of our cognition. Just start thinking and stop yourself the moment you token a thought with anything temporal in it: today, tomorrow, yesterday, the future, the past, earlier-than, later-than, a date, a temporal duration, a temporal order. How far did you get? Not far, we imagine. Now imagine you have the job of eliminating temporal thought altogether. That's a big job.

One *prima facie* cost of eliminativism, then, is that it involves a wide-scale elimination of a great deal of our everyday thoughts. Temporal thought is simply not like witch thought. Witch thought can, to a great extent, be isolated from most of our other thoughts, since we mostly don't and didn't think about witches. Temporal thought, by contrast, cannot be so easily isolated from thought in general. Much of our thinking may in fact employ temporal concepts. The widespread nature of temporal thought makes elimination not just difficult to achieve, but potentially quite damaging to our intellectual lives.

Now, to some extent, these worries for eliminativism rely on a naïve understanding of elimination. We can differentiate between *basic* and *sophisticated* forms of eliminativism. A basic form of eliminativism has it that we should eliminate some domain of discourse without seeking to replace it with anything. What the *prima facie* problems with eliminativism noted above reveal is that temporal thought serves a purpose and that we really want something to do the work that temporal thought does. Temporal thought has a role to play and we want something to play the role. A *sophisticated* form of eliminativism takes this moral to heart. Where appropriate, the sophisticated eliminativist doesn't simply recommend eliminating a domain of thought. She recommends, in addition, replacing that domain with something that can play the same role as the old domain, but that is strictly speaking true.

In order to get a feel for sophisticated eliminativism, consider a different example altogether. Consider again, the mereological nihilist we introduced in Chapter Nine. Mereological nihilists, recall, deny the existence of mereologically complex objects. As a consequence, they typically deny that a great deal of ordinary object talk is true. Strictly speaking, it is false to say that the cat is on the mat, assuming that both 'cat' and 'mat' refer to mereologically complex objects. There are no cats or mats. Rather, what exists, according to the mereological nihilist, are mereologically simple 'atoms': objects that have no proper parts. What looks like a cat on a mat is, in fact, a bunch of simples arranged cat-wise perched atop a bunch of simples arranged mat-wise.

Suppose that a mereological nihilist recommends *eliminating* our ordinary object talk (not, as far as we know, something that mereological nihilists do, in fact, recommend). A basic mereological nihilist will tell us to stop talking about cats and mats entirely. That would be disastrous for the cat—especially around feeding time. A sophisticated mereological nihilist, by contrast, recommends eliminating our ordinary object talk and *replacing it* with talk of simples. Thus, she says, speak not of cats; speak of simples arranged cat-wise. From the perspective of the simples arranged cat-wise, this strategy of replacing one kind of talk with another, rather than simply eliminating that talk and not replacing it, is much to be preferred.[2]

If there is some set of practices that we value, such that we can engage in those practices only if we have thoughts that play some particular role, we will say that there being thoughts that play that role is ineliminable relative to that practice. So, for instance, suppose we value certain practices around engaging

---

[2] One might argue that there's an even more sophisticated version of mereological nihilism. Rather than recommending that we replace our talk of complex objects with talk of mereological simples, we keep that talk but just recommend understanding it to be about something different than what we'd thought. Thus, talk of tables is *really* about simples arranged table-wise. Thunder (forthcoming) defends something pretty close to this view. The equivalent view of time would be that our temporal talk is not *really* about, say, a C-series. Rather, it is about something else, like an approximately spatiotemporal structure. This view is the non-aboutness realist position discussed in Chapter Nine.

with the things we call cats. We do not want to eliminate those practices, let us suppose, even if we were able to do so. It is plausible, however, that we could sustain those practices only if we had thoughts that play the role that our current cat thoughts do. So, relative to that set of practices, there being thoughts that play that role—thoughts that allow us to pick out the things we currently call cats—is ineliminable. But our having cat thoughts is not ineliminable, so long as some other thoughts can play that role. If having thoughts about simples arranged cat-wise can play the right role, then relative to those practices, cat thoughts are eliminable.

The temporal eliminativist, then, has to be committed either to the claim that there is no role that temporal thoughts play, that *needs* to be played given the practices we value, or to the claim that there is such a role, and something other than temporal thoughts can play that role. The latter strategy involves showing that we can replace temporal thoughts with some other thoughts, and that by doing so we are no worse off: the sorts of roles that the temporal thoughts played in scaffolding our prudential practices are now being played by the thoughts that replaced them.

As noted, we think it likely that there is a role that temporal thoughts need to play given the practices we value. In order to have agentive thoughts, something needs to play the role that temporal thoughts play. So, we think, sophisticated eliminativism is the only viable version of temporal eliminativism.

However, such a view faces three immediate difficulties. First, it is not clear that our temporal thoughts *can* be replaced. Our cognitive systems might be set up in such a way that we quite naturally employ temporal concepts across a wide range of cognitions. This might be quite heavily hard-wired. Even if it's in principle possible to replace these thoughts with others that do the same work for us, that doesn't mean that it's practically possible to do so; it certainly doesn't mean that it is feasible to do so.

Second, even if the replacement is in principle possible, doing so is a cost of some order. It would take time, training and effort to systematically replace temporal thought with something non-temporal. This is not a project that any fictionalist has to conduct. The fictionalist can leave all our ordinary temporal thought intact.

Third, it's not clear how much better off we would be by eliminating and then replacing our temporal thoughts. To be sure, if our temporal thoughts are systematically false then we would be epistemically better off by replacing those thoughts with thoughts that are not false. But if we replace temporal thoughts with thoughts that play essentially the same role as temporal thoughts in our cognitive economy, then we don't gain much conceptual advantage by doing so. We don't learn to see things in a fundamentally new way that gives us new capacities to engage with the world. Instead, we go about things in much the same way we ever have, but at the massive cost of the elimination and replacement of our temporal thoughts.

For that reason, at this stage of enquiry, it seems to us that eliminativism would be a costly direction to go in to pursue with little real benefit. Of course, we might turn out to be wrong about the benefits. Perhaps there would be considerable conceptual shifts that take place when we come to realise that time (in all of the folk senses) does not exist, and that our temporal thoughts and temporal talk are in error, and that the way we have been conceptualising the world is mistaken in certain crucial ways. Perhaps, in turn, shifting away from temporal thought would reveal new ways of thinking about the world, and our place in it, that would be fruitful to us. Perhaps it would even alter the way we see agency and deliberation. We certainly cannot rule this out: in part, that will depend on the details of the account of the thoughts we use to replace our temporal thoughts.

There is, however, a further problem for eliminativism, one that cuts deeper than the largely pragmatic costs identified above. To see the problem, put yourself in the position of someone who has just discovered that temporal error theory is true and has then been told that they should eliminate and potentially replace their temporal thoughts. Suppose that we put forward a three-step elimination program to help people reach a better, more fulfilling level of cognition. You need to decide whether to take the program and then act accordingly. But here's the rub: all your agentive thoughts are false! Why? Because your temporal thoughts are false and, as indicated above, if we assume a strong connection between temporal thought and agentive thought, it follows that your agentive thoughts are false too. So, it is unclear whether you can genuinely exercise your agency. If you can't, then you can't realistically *decide* to take up our three-step program and then *choose* to do so. Worse: you can't take up any program that leads to elimination, at least not via an act of choice, because you can't really choose to do anything.

An interesting question therefore arises at this point: can one genuinely exercise agency even though all of one's agentive thoughts are false? It is hard to see how. Consider the following thoughts about our three-step program: if I take the three-step program, then I will eliminate my temporal thoughts; if I take the three-step program, then my life will go better; if I don't take the three-step program then my life will go worse; if I take the three-step program, I will replace my temporal thoughts with something better. All these claims appear to have, embedded within them, temporal thoughts, particularly about what *will* happen if such and such an action is taken.

But all these claims are false. That's what temporal error theory tells us. Given this, it is very difficult to see in what sense you can genuinely bring it about, as a result of your active deliberation, that you no longer have temporal thoughts.

To put it another way, it's hard to see how you could have reasons to bring about the elimination of such thoughts. This is not to say that you cannot bring about such elimination at all. There are lots of ways we can intervene in our psychologies or cognitive states that do not go via conscious deliberation and choice. Drugs, certain therapies, physical damage to the brain, and so on, can all bring about, through causal interventions, certain changes to our cognitive states. So, it might very well be possible to bring it about that one eliminates one's temporal thoughts. But that is quite different from its being possible to deliberate about whether to do so, and to decide to do so on the basis of one's own reasons. So, even if it is possible to have agency in some cases without true agentive thoughts (and we remain sceptical), the kinds of agentive thoughts one would need to have in order to have reason to eliminate temporal thoughts seem to be the very thoughts that need to be true in order to be able to choose to act on the elimination.[3]

10.4. Temporal Fictionalism

Let us now consider fictionalism and see if the problems we have identified for eliminativism can be converted into benefits for fictionalism. The kind of temporal fictionalism we have in mind is the conjunction of three key claims. First, *an error theory claim:* our temporal thoughts are about time, and since time does not exist, our temporal thoughts are systematically false. Second, *a*

---

[3] Perhaps there is a difference between merely *having* agentive thoughts that are false and *having agentive thoughts while knowing that those thoughts are false*. It might be held then that there is only a problem for eliminativism if one both has false agentive thoughts and knows that those agentive thoughts are false. If one merely has false agentive thoughts but is ignorant of that fact, then perhaps one will at least be motivated to act in the ways needed to complete the eliminativist program. After all, one can, quite generally, be motivated to act by false beliefs. The trouble, however, is that it is hard to see how one could know that one's temporal thoughts are false while at the same time continuing to believe that one's agentive thoughts are true (or at least not realising that they are false), when they look like the thoughts above: namely, thoughts that have an embedded temporal structure. Perhaps this is possible for the unusually obtuse among us, but we fear that for most people it will be all too obvious that the relevant agentive thoughts are false if the temporal thoughts are, and so one won't in fact be motivated to act in a way that leads to the elimination and subsequent replacement of temporal thought.

fictionalist claim: our temporal thoughts are true in some appropriate temporal fiction. Third, *a normative claim*: we should engage in the temporal fiction. Things go better for us if we engage in the fiction than if we do not.

A temporal fiction is a story about the world according to which propositions such as 'it took 5 seconds for her to come up with a counterexample' are true in the fiction, but false otherwise. The easiest way to understand a temporal fiction is to imagine a complete description of a world in which time exists—the story that most of us take to be true—and now treat that description as the story of time. We can then talk about what's true in the temporal fiction by appending a fiction operator to the propositional content of temporal thoughts in the usual way.

We are not, however, principally interested in this kind of 'operator' fictionalism; we are concerned with how temporal thoughts connect to themselves and others, not merely in what sort of operator we could append to the propositional content of the thoughts to make them come out as true.[4] So, the fictionalism we have in mind is *pretence* fictionalism. According to pretence fictionalism, engaging in a fiction means adopting an attitude of make-believe or pretence toward a particular belief or package of beliefs.[5] That is, one pretends *as if* the relevant belief(s) are true.

Which package of beliefs? The beliefs whose contents are such that, when inside the scope of the relevant 'in the story of time' operator, those beliefs come out as true, and are false otherwise. According to pretence temporal fictionalism, one engages in a temporal fiction if one pretends that one's temporal thoughts are true. So, suppose Sara behaves excitedly today: she makes an utterance such as 'tomorrow is my birthday' and she makes arrangements for a party, and so on. Suppose, moreover, she believes that time does not exist and thus that, strictly speaking, there is no tomorrow. Thus, she takes 'tomorrow is my birthday' to be strictly speaking false. In this situation, Sara is engaging in the temporal fiction. She *pretends* as if her temporal thoughts are true.

The question before us, then, is whether the problems associated with eliminativism can be used to motivate the fictionalist alternative. In particular, can the difficulties at issue provide a reason to think that it is better for us to engage in a temporal fiction, and thus serve as a basis for the normative dimension of temporal fictionalism?

---

[4] For more on what we call 'operator' and 'pretence' fictionalism, and why in general to prefer the pretence iteration of the view, see Armour-Garb and Woodridge (2015: especially pp. 31–6).

[5] This view traces back at least as far as Walton (1990) who is one of its early advocates.

As we saw there are two broad difficulties for eliminativism: on the one hand, there is the pragmatic difficulty associated with elimination and, on the other hand, there is the worry that there is something self-defeating about eliminativism given the underlying temporal error theory. With respect to the first type of difficulty, the pragmatic difficulties for eliminativism arguably fall short of motivating the normative dimension of fictionalism. While it may be true that eliminating our temporal thoughts is costly, all that shows us is that we should continue to keep the temporal thoughts in question, even though they are false. It doesn't necessarily show that we should also behave as if they are true.

What of the worry that eliminativism is potentially self-defeating in virtue of the underlying temporal error theory? As we shall now see, far from motivating fictionalism, a version of the worry applies to temporal fictionalism as well. Ultimately, it appears as if there is simply no viable version of temporal error theory available if a strong connection between agentive and temporal thought is assumed.

10.5. Problems Revisited

Recall the basic problem. Temporal eliminativism seems to presuppose agency: in order to choose to eliminate those thoughts, and to act on those choices, one must have true agentive thoughts. One must deliberate about the process of elimination, choose to eliminate and thereby be motivated to act. However, if temporal error theory is true then one's agentive thoughts are false. In that kind of situation, it is very hard to see how anyone could reasonably choose to do anything, let alone eliminate one's temporal thoughts.

A similar difficulty arises in the fictionalist case. Suppose you learn that all your temporal thoughts are false. Then, according to the fictionalist, you should engage in a certain pretence: a pretence according to which you behave *as if* your temporal thoughts are true. However, one must *choose* to engage in the pretence. To see this, consider a different kind of pretence fiction altogether. Suppose that you are in a play. When you are in a play, you engage in some prop-oriented make-believe. You behave as if the broom handle is a sword, and as if being poked by the broom handle is a mortal wound. But it is clearly a choice that you make to behave in this way. The broom handle is not a sword, and you are not mortally wounded. But because you wish the play to go well, you behave accordingly. Similarly, in the case of temporal fictionalism, your temporal thoughts are not true. Nonetheless, because you want your life to go well, you choose to indulge in the pretence. You thus choose to pretend as if there is a future and as if there is a past.

The question is: how do you make this choice? To engage in the relevant pretence, you need to exercise your agency, but suppose that all of your agentive thoughts are false in virtue of the connection they bear to your temporal thoughts. So, it is false that if you choose to engage in the temporal fiction, life will go well. It is false that you can act in any particular way toward the future at all, because there is no future to act for.

In general, it seems you lack any prudential basis for acting one way rather than another. Here is why. Agentive thoughts, on this view, are false. Moreover, it's not just that in fact all the agentive thoughts we have are false. It's no mere accident. *Any* agentive thought we could have would be false. But, one might think, one can have normative reasons to do something—in this case prudential reason to continue to have agentive thoughts—only if one is an agent to begin with. Indeed, on various Humean views about normative reason, our normative reasons just are the reasons that we would have, given our motivating reasons, were we better informed and more rational. On views like that, if there are no motivating reasons because there is no agency, then there are no normative reasons, and hence no reasons of the right kind, to adopt fictionalism. At the very least, if agentive thoughts are false, one must tell some story about how we can have prudential reasons to engage in a temporal fiction, which does not appeal to our agentive status.

Now, if you could somehow get into the position of pretending as if your temporal thoughts are true then maybe, from there, you could pretend as if your agentive thoughts are true, and thus have some semblance of agency. But the trouble is that even this semblance of agency requires that you first have agency in some sense to get into the temporal fiction. Indeed, if anything your problems are doubled. For you must exercise your agency twice over: once to get into the temporal fiction and once to get into an agentive fiction. But you lack the prudential reasons that might motivate you to choose anything at all, and so it is not clear that you will exercise your agency once, let alone twice. To be clear, our claim is not that it is impossible for you to act. Nor is the problem that it is impossible to act as though your temporal thoughts are true. One can certainly imagine ways of stimulating or damaging the brain in such a way that it brings about exactly that pretence behaviour. The problem, at core, is a motivational one. If you believe that all your temporal and agentive thoughts are false, you lack any motivation to act in accordance with the normative demands of either temporal fictionalism or temporal eliminativism. So, you won't eliminate your temporal thoughts, and you won't behave as if they are true either. You will just continue to have them while believing that they are false and behave accordingly. We say 'behave accordingly' but we doubt that you will be motivated to behave *at all*. While you may have the capacity to act, in some sense, you won't ever be motivated to do so, and not just when it comes to the normative demands of eliminativism and fictionalism. You won't, realistically, even be motivated to choose to get out of bed in the morning.

The situation starts to look worrying indeed. We can now see that eliminativism and fictionalism are options for managing the dangerous outcome of believing that all your temporal thoughts are false. If one can replace one's temporal thoughts with something better, or behave as if those thoughts are true, then one has at least a *chance*—slim as it may be—of keeping the motivational and thus the normative structure of agency intact. If one is forced to keep temporal thought as it is and cannot choose to pretend as if it is true, then the potentially radical implications of temporal error theory are realized. Temporal error theory seems to undermine agency.

At this point it is perhaps useful to draw a distinction between two kinds of fiction: an *optional* fiction and a *forced* fiction for it may appear that pursuing the route of treating the temporal dimension of our behaviour as a *forced* fiction generates a natural response to the worry voiced above (though as we shall see, it does not). An optional fiction is a fiction that we have the option of choosing to engage with. The fiction of a play is like this. One can choose to behave as if a broom handle is a sword, but equally one can choose to behave as if the broom handle is just a broom handle and thus flatly refuse to engage in the prop-oriented make-believe at issue. A forced fiction, by contrast, is a fiction in which one cannot and does not choose to engage. One engages in that fiction *because one has to*. Thinking about *The Truman Show* gives us some sense of what a forced fiction might be like. A great many of Truman's beliefs about the world he lives in are false. Moreover, everyone around him is engaging in a complex game of make-believe in a careful attempt to keep the fictional world that Truman lives in alive. But it is not as though Truman ever *chose* to behave as if his beliefs about the world are true. The fiction, in a certain sense, was *forced upon him*. Perhaps the temporal fiction is like this: it is a fiction that we are, in a sense, forced to engage in.

The analogy to *The Truman Show* also reveals the limitation of thinking of temporal fictionalism as a *forced* fiction. At a certain point, Truman discovers that most of his beliefs are false, and that he lives in a world of make-believe. At that point, he faces a choice: either he can choose to continue living his life, pretending as if his beliefs are true, or he can choose instead to behave as if those beliefs are false. In the end, he chooses to exit the fiction. The problem is that we are in the same position as Truman: we come to learn that our temporal thoughts are false and at that point must make a choice. Either we continue to behave as if our temporal thoughts are true or we escape the fiction. In a certain sense, then, the fiction is forced in exactly the same way as Truman's fiction was, but this turns out not to be enough. It is not enough, because at a certain point we must still face the choice of whether to behave as if our temporal thoughts are true. At that point, we face the kind of agentive paralysis that temporal error theory seems to generate.

And so, the thought goes, perhaps this signals a way for the temporal fictionalist to respond. Let us imagine that Truman discovers that all of his beliefs are false, but he *also* discovers that if he doesn't behave *as if* his beliefs are true, and thus continue to lead his life on the show, he will be euthanised and replaced with a new star. In that situation, there is a sense in which Truman can choose, but also a sense in which he doesn't *genuinely* have a choice about whether to stop engaging in the fiction. What if the temporal fiction is more like this? One way this might happen, we suppose, is if behaving as if our temporal thoughts are true is necessary for agency, and agency itself is necessary for being a person, as hinted at above. In this situation, the choice about whether to continue behaving as if our temporal thoughts are true is not a genuine choice, in the way that Truman's isn't when he learns about the threat of euthanasia. For if we choose not to behave as if our temporal thoughts are true, we are effectively choosing to not be persons anymore. In a slogan: give me death or give me time!

Of course, in a certain sense, it is not true that Truman has no choice. His choice is *unfair*, his choice, we may think, is morally problematic in all sorts of ways. But, one might argue, the bottom line is that he still has a choice and, indeed, *must choose*.[6] He cannot simply choose to do 'nothing' as it were, because that, in itself, is a choice; it is a choice to behave as if his beliefs about the world around him are true. So too in the case of fictionalism. Discovering that one's temporal thoughts are, in general, false in a certain sense forces a choice: one must either choose to behave as if those thoughts are true or choose to behave as if the thoughts are false. The choice may be unfair, and problematic in all sorts of ways, but one cannot simply refuse to do anything. For that, in a sense, is to make a choice: it is to choose to engage in the pretence.

---

[6] Even if Truman doesn't have a rational reason to choose, he may still have a purely pragmatic reason to do so. It may be that Truman is in a Pascal's wager-style situation, where there is no rational basis to make a choice, but there are nonetheless significant pragmatic benefits to preserving the status quo.

Perhaps, however, the case of temporal fictionalism, and the imagined twist on *The Truman Show* are not totally analogous. The way the fiction is forced in the temporal case, one might argue, is not the same as in Truman's case. Truman's engaging in the fiction is a result of some choice. But perhaps not all forced fictions are like this. It may simply be that, due to features of someone's psychology, or due to certain causal facts, one cannot stop holding certain beliefs, but where this is not really a matter of choice. It is just that one cannot divest one's self of the relevant beliefs. The difference, then, with the Truman case is that while Truman is in a position to choose what to believe, the temporal fictionalist is not, but must proceed with her false beliefs anyway.

Perhaps a version of temporal fictionalism along these lines is indeed viable, though we daresay it would require some spelling out: what are the factors that prevent one from holding certain beliefs? How can these factors undermine the capacity to choose in the manner required to strip the forced fiction of any whiff of agency? Ultimately, however, we won't pursue this line of thought, since we believe that there is a way to make sense of choosing to engage in a temporal fiction, to be described in the last chapter.

10.6. Conclusion

In this chapter, we have considered temporal error theory in its two guises: temporal eliminativism and temporal fictionalism. Ultimately, what we have shown is that the prospects for both views are intimately tied to how we think about the relationship between agentive and temporal thoughts. Everything we have said here thus relies on thinking about the relationship between agency and time in a certain way. We believe, however, that we should take the problems that arise from temporal error theory as an indication that the relationship between agency and time must be reconsidered. Put the point this way: if there is any way to make agentive thoughts true *even though* temporal thoughts are false, then there would be scope to engage in the kind of wilful behaviour needed to behave as if our temporal thoughts are true.

Thus, in the final chapter, we turn to agency and consider it in some detail. We show how to recover a realist picture of agency in the face of temporal error theory. We then use our case for agentive realism to make a case for temporal fictionalism over the eliminativist alternative. The picture we present shows how to achieve a combination of temporal error theory in its fictionalist guise, and agentive realism.

11

Time and Agency

11.1. Introduction

In the previous chapter, we identified a problem for temporal error theory. There are, broadly, two forms of temporal error theory available: temporal fictionalism and temporal eliminativism. Both forms of error theory impose a normative demand. Eliminativism demands that we eliminate our false temporal thoughts; fictionalism demands that we behave as if our temporal thoughts are true. The trouble is that it is hard to see how these normative demands can be met.[1] In particular, it is unclear how anyone could be motivated to act in one way or another, so long as we assume that our temporal thoughts and our agentive thoughts are linked, in the following sense: if one's temporal thoughts are false, then one's agentive thoughts are false.

In this chapter, we aim to develop a realist approach to agency. The key is to reconsider the link between agency and time. We suggest that agentive thoughts can be based on non-temporal foundations, allowing for a viable notion of agency in the face of temporal error theory. The picture of agency we develop still invokes temporal thoughts in some cases, and so the link between time and agency is not completely severed. It is, however, weakened to the point where it is possible to gain enough of an agentive foothold to make temporal fictionalism viable.

The chapter begins, in §11.2, by situating our discussion within a framework of realist and anti-realist accounts of agency. Then, in §11.3, we outline two arguments for the conclusion that our agentive thoughts (in particular,

---

[1] One might reject the claim that both forms of error theory impose a normative demand. For instance, one might claim that a view on which the fiction is forced (i.e., one in which we can't help but hold false beliefs) imposes no normative demands. Alternatively, one might argue that a view in which we take our temporal thought to be about something other than what we originally took it to be about imposes no normative demands either. We disagree. The first view still imposes normative demands; it is just that those demands are satisfied by default. Just because a demand is met, even by default, does not mean that there is no such demand. The second view, by contrast, is best thought of as a form of realism (namely, non-aboutness or revisionary hermeneutic realism). While this view does not impose normative demands, it is not a version of error theory, and our claim only applies to error theories about temporal thought.

Out of Time: A Philosophical Study of Timelessness. Sam Baron, Kristie Miller, and Jonathan Tallant, Oxford University Press. © Sam Baron, Kristie Miller, and Jonathan Tallant 2022. DOI: 10.1093/oso/9780192864888.003.0011

some of those we would have taken to be true) are false if temporal error theory is true. We subsequently (§11.4) focus on anti-realist views of agency: those that accept the conclusions of both arguments outlined in §11.3. We argue that anti-realism about agency is unattractive, which leads us into our discussion of realism (§11.5) where we defend a broadly realist picture of agency and agentive thought, one that nonetheless involves a limited degree of fictionalism: limited because it is fictionalism about only *some* agentive thoughts. As we shall see, this limited fictionalism about agentive thoughts not only relies on the temporal fictionalist position discussed in Chapter Ten, it also supports it. In §11.6 we consider a difficulty for our account of agency, namely that a similar project can be used to recover folk notions of time.

11.2. Agentive Concepts and Agentive Thought

Let us begin with agentive concepts. We have a concept of agency: of what it is to be the sort of thing that reasons, deliberates, intends, and acts. We thus also have concepts of reason and deliberation; and of planning, intending, and acting. There is an entire web of agentive concepts that includes all these concepts and more. Agentive concepts are legion, and we don't intend to mention them all. Rather, we just want to gesture towards the phenomena we have in mind.

To be clear, though, when we talk of our agentive concepts we don't have in mind a theoretically laden view of these concepts. We don't have in mind any particular philosophical account of reason, intention, action, deliberation, or agency. Rather, we have in mind the sorts of concepts that we employ in everyday life as we navigate the world. These are the concepts we employ when we try to figure out what we should do. They are the concepts that, in some good sense, shape our engagement with the world. Insofar as we engage with the world as though we are intenders and actors, we engage with the world by intending some things and not others, by acting in some ways and not others. These are the concepts we employ when we assume that we are not merely passive in the world, that we are not simply acted upon, but that we also act upon things. These are the concepts we employ when we assume that we can intervene in the world in ways that make things better (and often also worse) for us.

Agentive thoughts are thoughts that employ agentive concepts. We thus include any thoughts about what one can, or should, do, either morally or prudentially. So, for instance, when Freddie intends to catch the kangaroo by running a little faster, or when Annie thinks that she should buy Jane a present for her birthday, or when Julie the jolly cannibal deliberates about whether or not to eat her very noisy neighbour, these all count as agentive thoughts.

In what follows, we suppose there to be a connection between agentive thought and agentive concepts. Namely, we assume that our agentive thoughts are in good standing if and only if our agentive concepts are (at least by and large) satisfied. For now, let us suppose that to say thoughts are in 'good standing' means that by and large those thoughts are true—so we set aside the view that they are not truth-apt at all at this stage. So, in what follows we will move between talking about whether certain of our agentive concepts are satisfied and talking about whether our agentive thoughts are true. We make the assumptions that license this vacillation to be relatively uncontroversial: for instance, it's hard to see how our agentive concepts could be satisfied while our agentive thoughts are in error, or, conversely, how our agentive thoughts could be true, and yet our agentive concepts be unsatisfied. This is, so far, the very same sort of reasoning we adduced in Chapter Nine when we argued that our temporal thoughts are true just in case our temporal concepts are satisfied.

Using the same classificatory scheme that we applied to temporal thought in Chapter Nine, we can differentiate between a range of different accounts of agentive thought.

Agentive Realism: ordinary agentive thought is truth-apt, and many core thoughts in that domain are true.

Agentive Error Theory: ordinary agentive thought is truth-apt and core thoughts in that domain are not true.

Agentive Non-Cognitivism: core thoughts in the agentive domain are not truth-apt because such thoughts are not cognitive.

As before, within error theory there are two further positions: fictionalism and eliminativism.

Agentive Eliminativism: (i) ordinary agentive thought is truth-apt; (ii) core thoughts in the agentive domain are not true, and (iii) ordinary agentive thought ought to be eliminated.

Agentive Fictionalism: (i) ordinary agentive thought is truth-apt; (ii) core thoughts in the agentive domain are not true, but nevertheless, (iii) ordinary agentive thought constitutes a useful fiction, and one that we should continue to engage with.

In Chapter Ten, we assumed that if temporal error theory is true, then agentive error theory is also true. In the following section, we consider what we take to be the two arguments available for this connection: the cognitive argument, which focuses on the relationship between agentive and temporal thoughts, and the conceptual argument, which focuses on the relationship between agentive and temporal concepts, with attendant implications for agentive thought. Both arguments assume that folk conceptions of time are in error, and that as a result our temporal thoughts are false (thus leading to some form of temporal error theory). This second assumption might seem odd, given the difficulties that we have outlined for temporal error theory in the previous chapter. However, as we also argued in Chapter Ten temporal error theory seems to be the right view, given that our folk concepts of time go unsatisfied. Moreover, the problems for temporal error theory stem from its potential connection to agency, and so we don't see any problem with assuming the view for the sake of argument here, since it is that connection that we aim to undermine.

11.3. Two Arguments for Agentive Error Theory

Let us begin with the cognitive argument for agentive error theory. This argument proceeds as follows. Some of our agentive thoughts explicitly token temporal thoughts. For instance, consider the thought that in order to get to the store by 5.00 p.m. Harry must leave by 4.00, or the thought that Hermione's birthday is on July 1, and so Ron needs to make sure to buy her a present before June 28; or the thought that it will take Hagrid 10 minutes to get from place A to place B if he walks, and 20 minutes if he drives his motorcycle, and so he should walk.[2]

Some agentive thoughts are *time-involving*: they feature embedded temporal thoughts. If temporal error theory is true, however, these embedded temporal thoughts are false. If the embedded temporal thoughts are false, however, then the agentive thought itself is false. So, for instance, if it is false that Hermione's birthday is on July 1 because there is no C-series ordering that can support the temporal attribution that birthdates require, then it is also false that Ron needs to buy Hermione a present before June 28 in order to ensure she has a happy day.

---

[2] Traffic around Hogsmead is clearly pretty bad.

So ends the cognitive argument. The cognitive argument may end up undermining all our agentive thoughts. For one might think it is plausible that all agentive thoughts involve the tokening of temporal thoughts, they just do it in a way that is less obvious. Consider a thought of the form 'I will/could F'. You might think that for Freddie to token the thought 'I will/could F' is for Freddie to token a thought that involves (tacitly or otherwise) tokening the thought that there is some time at which he might or does F. But since Freddie's thought that there is a time (any time) is false, it follows that his agentive thought is also false. If all agentive thoughts are of the above form, then the cognitive argument for agentive error theory is far reaching indeed. Still, the cognitive argument potentially leaves it open that there are some agentive thoughts that are true if temporal error theory is true. For that argument will not apply to any agentive thought that does not token a temporal thought. That being so, if the cognitive argument succeeds this may not spell any kind of disaster for agency in general. At least, that is so, as long as enough agentive thoughts escape the argument. More on this later.

This brings us to the second argument for agentive error theory: the conceptual argument. The conceptual argument threatens to undermine all agentive thoughts, regardless of whether they token temporal thoughts. The basic idea behind the conceptual argument is this: accepting that our temporal concepts are in error undermines the core concept of agency. If the core concept of agency is in error, however, then agentive thought is generally in error too. From this it follows that agentive thought is systematically false and that agentive error theory is true.

The conceptual argument essentially relies on the assumption that when our temporal concepts go unsatisfied, this undercuts the core concept of agency as well. To see why this assumption might be plausible, let us consider what the concept of agency in question might be. Here's one possibility:

x has agency only if (i) for at least one φ, x can deliberate about whether to φ and (ii) x can φ as a result of that deliberation.[3]

---

[3] This is an analysis of a very thin concept of agency, one that requires only the capacity to deliberate about, and subsequently act upon, at least one thing. A thicker notion of agency may require there to be a set of things one can deliberate about and act upon. We can easily modify the above conceptual analysis to capture this thicker notion, as follows:

x has agency only if (i) for some sufficiently large set y, x can deliberate about whether to φ for each member φ in y and (ii) x can φ as a result of that deliberation.

Everything we say below can be said using this thicker notion of agency instead if necessary.

We will say a bit about the status of this conceptual analysis in a moment. First, however, let us consider how this concept might be used to formulate the conceptual argument for agentive error theory.

One way to see this is via the notion of deliberation. It seems to be essential for anyone to engage in deliberation that they take some things to be fixed, and unchangeable, and to be the things on the basis of which they deliberate. Moreover, it also seems to be essential that they take other things to be open, and malleable, and to be things about which they deliberate. Further, it seems essential to suppose that some ways of acting will have downstream consequences, and hence that it makes sense to deliberate about what to do, in order to bring about certain ends. It is not possible to deliberate about something one takes to be fixed and unchangeable.[4] One cannot typically deliberate about whether the sun exists tomorrow. It is also not possible to deliberate about something one takes to be open and malleable, unless one also takes other things to be fixed. For in the absence of there being some fixed things, and hence some fixed pattern of events, one has no basis on which to suppose that certain actions will have these, rather than those, consequences. If one knows nothing about the Humean mosaic (if you will) then one has no more reason to think that eating one's morning porridge will satiate one's hunger, than to think it will cause one to levitate, or spontaneously explode.

In what follows we will suppose that deliberators must *impose* something like this basic structure on the world if they are to deliberate. We will call the structure that they must impose, the *deliberative structure*. We hold some things fixed; we hold some things open. It does not follow from this that some things really are fixed, and others really are open (at least in a metaphysical sense) as, for instance, they would be if our world were a growing block world.

Still, it seems clear that in order to impose a deliberative structure, one must believe that our world has a structure that supports that imposition. We will say that a 'supportive structure' is any structure that one takes to obtain in our world and which supports individuals imposing a deliberative structure. To help understand the notion, suppose for a moment that there is time, and that our world is a growing block world. Then, our world's being a growing block world in which we have epistemic access to (some) facts about the objective past, but not access to (all or most) facts about the future (since there are no such facts) is one way our world could have a supportive structure. But it is not the only way: our world could be a block universe world that has features such

---

[4] Price (2007) makes this point.

that we have epistemic access to past times, but not future ones (or some such). Then this, too, would be (part of) a supportive structure.

Now, suppose it were to turn out that the presence of a C-series is a necessary component of any supportive structure. Thus, if our folk temporal concepts are in error, then there is no supportive structure. If one thereby accepts this fact, then one cannot impose a deliberative structure on the world, because one will not believe there to be sufficient supportive structure to do so. Thus, one cannot deliberate about whether to φ. From this it follows that the concept of an agent outlined above is in error. If this concept or any concept like it underwrites one's agentive thoughts, however, then one's agentive thoughts will fail too.

Thus stated, the argument only establishes that one's agentive thoughts are false *as soon as one comes to believe that there is no C-series*. It is thus only the case that if everyone believes this fact, that agentive error theory follows as a general thesis. Nonetheless, the fact that believing that the C-series does not exist renders one's agentive thoughts false is bad enough. For if we are compelled to accept that there is no C-series based on discoveries in science, then we are forced to give up agency as well. As a corollary, temporal error theory itself becomes unstable. If accepting that there is no C-series compels one to adopt temporal error theory, and accepting that there is no C-series also means that one cannot thereby be an agent (because one does not satisfy the concept of agency), then the problems identified for temporal error theory in the previous chapter are realised. For, as argued there, the viability of temporal error theory hangs on one's capacity to exercise agency.

As we have stated it, the conceptual argument relies on a certain conception of agency. Let us now return to say a bit more about the conceptual analysis of agency proposed above. What is the status of this conceptual analysis? Is it supposed to capture the folk concept of agency? Determining whether, and to what extent, this is so, would require a thorough empirical investigation of our folk concept or, indeed, concepts of agency. It would, at a minimum, require something analogous to our investigation of the folk concept(s) of time. Such an investigation would tell us whether there is a single shared, concept of agency and go at least some way towards telling us under what conditions such a concept is satisfied. Since we are particularly interested in the connection between agency and time, such an investigation would ideally tell us about the ways in which the concept(s), of agency, is connected to the concept(s) of time. Unfortunately, we do not have these empirical resources to draw upon in this chapter: research of the right kind into the folk notion of agency simply does not exist.

Nonetheless, we are going to assume the conception stated above in what follows. This might seem more than a little awkward for us, given our stated naturalism in the opening chapter of this book. There we stated that any discussion of concepts ought to be conducted in an empirically sensitive manner. We continue to uphold this commitment. We believe, however, that our stated naturalism allows the making of empirical bets in the absence of evidence to the contrary. We may take a bet on what the world is like, so long as the bet is clearly flagged as such. We are thus willing to bet that people's actual concepts of agency are something like the concept outlined above. Note that in Chapter Two, we set aside similar bets about the folk notion of time (such that it would continue to be satisfied despite what people report in experimental conditions). We will return to say a bit more about why we're willing to make one bet and not another later. First, we will take a closer look at agentive error theory, before offering our realist alternative.

11.4. Against Agentive Error Theory

Before we provide a response to the cognitive and conceptual arguments for agentive error theory, let us first motivate the need to provide a response at all: what's so bad about agentive error theory? As noted, there are two versions of the view: eliminativism and fictionalism. Making the case against agentive eliminativism is straightforward. After all, this is the quite radical view that we should eliminate agentive thought. It is almost impossible to imagine how one could accomplish such a feat, and, if we did, what sort of creatures would remain.

This suggests that the agentive error theorist would be better off embracing agentive fictionalism. We take it that the most natural version of agentive fictionalism would be a version of pretence fictionalism. We could think of the pretence as being the pretence that our agentive thoughts are (by and large) true.

We briefly considered a view along these lines in the previous chapter, when we considered temporal fictionalism. The main problem noted there is that it is simply unclear how to make sense of the idea of acting (both in thought, talk, and behaviour) *as though* one is an agent. We are supposed to be, as it were, acting *as though* we are agents.

But the idea of acting as though one is an agent is rather different from the idea of acting as though Santa Claus exists, or as though there are objective moral truths. For how does one go about acting *as though one is acting*?

Action, and so acting, is itself agentive. So is intending and intention. So how does one act *as though* one is an agent? If one were genuinely acting, then surely one *would* just be a genuine agent. So, whatever one is doing, it is not that. Or to put the point the other way around: imagine someone tells you to act *as though* you are an agent, even though *in fact you are not*. It's not clear what advice you are being given. At best, it seems, you should reply that you can act* as though you are an agent. But you cannot act as though you are an agent, because acting is something that only agents can do. If that is right, though, then all the fictionalist has done is move the bump in the carpet: because what exactly is an act*?[5]

Even worse: *which* fiction should we engage with? Suppose Jeremy is discussing with Sue whether or not he should eat more—rather than fewer—muffins in his bid to lose weight. Sue says that he should eat fewer muffins in order to lose weight. Jeremy says that he should eat more muffins in order to lose weight. We have before us, then, two agentive thoughts that Jeremy might entertain. Which of these does he pretend is true? We want to say, relative to his goal of losing weight, that what Sue says seems to be right (except in the rare situation in which Jeremy replaces eating tubs of lard with eating the same number—but not mass—of muffins).

However, it is difficult to make that claim without appealing to a bedrock of true temporal thoughts. What we *want* to say is that it is just true that Jeremy *will* lose weight in the future if he takes one action, and false otherwise. Given the underlying temporal error theory, according to which all such thoughts are false, it is unclear that one fiction can be held up as any better than the other. Indeed, we cannot say that engaging in one fiction will turn out better for Jeremy and so he should engage in that fiction, because that in itself is an agentive thought, and one that is false on the current picture.[6]

We have framed this problem as a difficulty for fictionalism, but a version of it applies to eliminativism as well. If you are like us, then your initial reaction to agentive eliminativism was probably to think that such a view is not merely

---

[5] Moreover, even if we could make sense of acting*, it's not clear that agentive fictionalism would have the sort of bite that we would hope. Consider moral fictionalists. They argue that we have prudential reasons to engage in a moral fiction. Indeed, presumably the sorts of prudential reasons to which they appeal are ones that give us reason to engage in particular sorts of moral fiction: a fiction, for instance, in which it is not considered morally right to eat one's neighbour if they play their music too loud. A moral fiction on which 'anything goes' would not, we take it, be one that, were we to engage in it, would furnish the sorts of benefits that moral fictionalism promises.

[6] The intuition that some agentive thoughts are true, and others not, and that we can criticise (at least sometimes) those with false agentive thoughts, is very appealing. We don't say that there is no way for the agentive fictionalist to respond to this worry. Still, the problem strikes us as sufficiently severe to impede the development of a plausible version of the view.

radical, but quite mad. How could it possibly be a good idea to give up on agentive thought? Indeed, when discussing eliminativism we noted that to do so would be to *give up* on many of the things that we take to be central to us being the very kinds of things we are. On second thought, though, this seems like an odd way to put matters if we are taking agentive error theory seriously. After all, agentive error theory is the view that our agentive thought is systematically in error. But to talk of us *giving up* many of the things that we take to be central to us being the very kinds of things we are, makes it sound as though, as things stand, we are agents, and we would be giving up being agents were we to eliminate our agentive thoughts. But of course, that cannot be right if agentive error theory is true. It's not as though prior to the elimination of witch-thought there really were witches, and that by giving up those thoughts people gave up magic. Likewise, if agentive error theory is true, then we are not agents. Giving up agentive thought is not giving up anything: it's simply ceasing to engage in a way of thinking that is systematically false. Put like that, though, it seems clear where agentive eliminativism goes awry. It's not that each of us is committed to thinking that we should continue to have agentive thoughts even if in fact we are not agents. Rather, it is that we are committed to us *in fact being agents*.

That, in turn, brings us back to what we take to be the central problem for agentive fictionalism: that the very idea of pretending to be agents seem to require that we are, already, agents. Pretending is the sort of the thing that, surely, only an agent can do. It's no accident that footstools don't pretend to be cats. Pretending requires intending, and planning, and acting. And these are things that require, or perhaps better, are constitutive of, agency. All of this motivates the need for agentive realism.

11.5. Agentive Realism

In what follows we cannot hope to spell out a full-blown realist account of agency without time. That would be a massive undertaking. Instead, we will focus on showing how the realist can respond to the two arguments we mounted earlier: the conceptual and cognitive arguments. That will require spelling out some aspects of a timeless account of agency, but it won't be anything like a complete account.

Let us start with the conceptual argument. Recall that this argument moves from a particular conception of agency coupled with a belief in temporal error theory to the conclusion that one's agentive thoughts are false. The concept of agency at issue is one that focuses on deliberation. As already discussed, we believe it is plausible that for one to be an agent, one must be able to impose a deliberative structure. We are also inclined to accept that in order to do that, one must believe that there is some supportive structure. We nonetheless think the conceptual argument fails. It's not the case that there must be a C-series for there to be a supportive structure. Thus, one can reasonably believe that there is a supportive structure even if one accepts temporal error theory. There is thus no bar to the imposition of the kind of deliberative structure needed for agency. In what follows we outline how a supportive structure can be provided by the presence of one kind of structure in particular: namely, a causal structure.

Consider the idea that we need to take some things to be fixed, and unchangeable, and to be the things on the basis of which we deliberate. Consider also the idea that we take other things to be open, and malleable, and to be things about which we deliberate. Causal structure seems well suited to support both aspects of deliberative structure. There being a causal structure allows us to take as fixed, and unchangeable, those things that are *causally prior to* our location, and as open and malleable, only those things that are *causally downstream* from us. We cannot causally intervene on events that are causally prior to us, so they are excellent candidates to be held fixed and unchangeable.

By contrast, and at least in principle, we can causally intervene in events that are causally downstream from us, so these are also excellent candidates to be held open and malleable. Indeed, with respect to those things that are causally downstream from us we can distinguish between those that we take ourselves to be able to causally affect, and those we take ourselves not to be able to causally affect. Only those in the former category are those about which we can deliberate. Those events that are causally prior to us, in the causal ordering, or those that are causally downstream from us, but which we take ourselves to be unable to causally affect, are those things about which we cannot deliberate. So, the presence of causal structure provides a supportive structure for at least these aspects of deliberative structure.

In addition, the presence of causal structure supports our imposing a deliberative structure that allows us to deliberate on the basis of those things that we hold fixed. Very plausibly, we can take those events that are causally prior to us as the basis of our deliberations. We see a particular *causal* pattern in those events, causally prior to ourselves. On the basis of that pattern, we infer that certain actions will have certain downstream effects. It is because, for instance, one sees that there is a causal connection between eating porridge and feeling satiated, and no such pattern between eating porridge and exploding, that one can sensibly decide, on the basis of deliberation, to eat porridge for breakfast in order to satiate hunger.

The use of causal structure to support agency is not particularly radical. What it is to be an agent is, at least in part, to be the kind of thing that can deliberate about what to do, and who, on the basis of that deliberation, can causally intervene in the world. Intentions are intentions to causally intervene, and actions are the procedure of such interventions. So, one would expect an account of agency to be deeply connected to an account of causation. In turn, it should be no surprise that the presence of causal structure can play the role of supporting deliberative structure. It may even be that not only can causal structure be, or be part of, the supportive structure, but causal structure is a necessary component of supportive structure. That strikes us as a plausible claim, but not one we argue for here, and not one that we need in order to defend agency from the conceptual argument.

Our response to the conceptual argument, then, is just to note that time in the folk sense is not necessary for there to be deliberative structure. If time in the relevant sense were necessary for causation, then temporal structure would be necessary for supportive structure. Once we sever any necessary connection between time and causation, however, it becomes clear that temporal structure is not a necessary component of supportive structure.[7] That being so, it is possible for an agent to impose deliberative structure on the world even though they believe there is no C-series. They can do this so long as they continue to believe in the existence of causation. If that's right, then an agent is never forced into the awkward position of undermining their own agency by accepting the implications of a scientific theory that eliminates the C-series. Matters would be different, of course, if a scientific theory were to show that causation does not exist. But, as discussed in Part Two of the book, a situation in which a scientific theory that eliminates the C-series is confirmed is also likely to be one in which causation exists as the basis for confirmation. As far as we can tell, the conceptual argument is undermined by the very fact that enough structure remains, even in the absence of time, for empirical confirmation.

---

[7] Of course, in order to fully satisfy the concept of agency outlined in the previous section, it must also be the case that one can act on the basis of one's deliberations. But so long as there is causation, we see no reason why one cannot perform an action. The presence of a C-series, so far as we can tell, is not necessary for one to do anything, except in so far as the C-series is needed for causation (which we have argued it is not).

This brings us to the cognitive argument. Recall the basic argument: some if not all of our agentive thoughts involve temporal thoughts. Those thoughts are therefore true only if the relevant temporal thoughts are true as well. If temporal error theory is true, then our temporal thoughts are false and so at least some agentive thoughts are false too. Our response to this argument is just to concede the conclusion: yes, some of our agentive thoughts are false. Be that as it may, the falsity of the relevant agentive thoughts does not undermine agency. The reason for this is that there are other nearby agentive thoughts that are true; thoughts that invoke only causal notions.

Consider, for instance, Harry's agentive thought that he needs to get to the store in the next five minutes to get Hermione a present. While it is not strictly speaking true that Harry has to get to the store in five minutes to get Hermione a present, what is true is that Harry's getting Hermione a present is causally dependent on Harry's making it to the store, and, moreover, that if Harry gets to the store and the store is shut, then Harry cannot get the present. If we replace Harry's false time-involving agentive thought with a causation-involving agentive thought, then we end up with a thought that can play the right kind of role in Harry's life. It motivates him in the right way, allows him to appropriately achieve his desires, and it is a thought that can be the subject of deliberation. So far as we can tell, there is no functional difference for Harry between having the time-involving agentive thought and having the nearby causation-involving agentive thought.

One might disagree with this diagnosis. Harry needs to believe that he must get to the store within five minutes, otherwise he won't be appropriately motivated to get off the couch. But there is no nearby causal thought, one might argue, that is strictly true and that can provide such a *pressing* motivation for action. In fact, though, there is a nearby causal thought that can do this work, it is just quite complicated. Consider Harry's clock. There is a causal story about what the clock shows on its face in terms of the mechanical workings within the clock. In principle, one can transform the claim that the clock shows five minutes have passed into a causal story about the mechanism inside the clock. Doing so is not straightforward, but it is not impossible. Harry can come to believe that if the clock is caused to show a certain number, then it won't be possible to buy Hermione a present. That belief seems like the right kind of thing to motivate Harry into action, and it seems just as pressing as the associated temporal thought. It is, we submit, this more detailed causal story that can play the same role in Harry's life as thoughts about time.

The shift to causation here plays much the same role as the shift to causation that we recommended in response to the conceptual argument.

The shift allows us to keep agency despite the truth of temporal error theory, on the assumption that the C-series is not necessary for causation. Of course, the causal thoughts may be much more complex, and so the shift to causation increases the cognitive load of agency, a point we return to in a moment.

Still, if we are right that time-involving agentive thoughts can be converted into ones that involve only causal thoughts, then the cognitive argument can be easily avoided. This raises the question, however, as to *why* we should keep the strictly false agentive thoughts at issue around. Why not simply eliminate the time-involving agentive thoughts and replace them entirely with ones that involve only causation?

The answer is that the temporal thoughts are much *simpler*. As noted, Harry's thought that he needs to get to the shop in five minutes to get Hermione a present could be replaced with a complex causal thought involving the inner workings of Harry's clock. But it is not in Harry's best interest to have this complex causal thought. It is much more efficient for him to have the (admittedly false) temporal thought that he must leave in five minutes so he can get out the door. Harry's time-involving agentive thoughts are a cognitively efficient way of compressing true causal information.

Note that we are not making the controversial claim that time can be reduced to causation or *vice versa*. Indeed, we are saying that while the time-involving agentive thought is an efficient proxy for the causal thought, it is still a distortion of reality, which wouldn't be the case if time and causation stood in some reductive relationship. The idea, rather, is that causation is extremely complex, and giving a total causal story is sometimes out of reach. We can see time-involving agentive thoughts as a way of describing real causal structures in idealized (because false) temporal terms, and this is useful because it eases cognitive load.

Indeed, our view is that the standard time-involving agentive thoughts, while false, can do all the work of the nearby true causal thoughts. That being so, we have good reason not to eliminate them, even though they are false. One might disagree with this. Recall Jeremy's thought that if he wants to lose weight, he should eat more muffins in the future. There appears to be all the difference in the world between Harry's thought that he needs to get to the store within five minutes in order to get Hermione's birthday present, and Jeremy's thought. Harry's thought is apt, in a way that Jeremy's thought is not; Harry's thought is useful in a way that Jeremy's thought is not and so on. If we are to continue using our false time-involving agentive thoughts, then this is a difference that must be accounted for, and it cannot be accounted for in the obvious way: by pointing out that Jeremy's thought is false. Both thoughts are false.

Can the difference between Harry's thought and Jeremy's thought be recovered on the view we are proposing? It can. While Harry's time-involving agentive thought is false, we can replace it with a true causal thought in the neighbourhood. Things are much worse for Jeremy. Jeremy's agentive thought is false, and there are no true causal thoughts nearby. It is just not true that eating muffins is causally related to weight loss. That's why Jeremy's thought fails to be apt, or useful or whatever.

The view we are proposing is a *limited fictionalism about agency*. Many agentive thoughts are straightforwardly true. These are the ones that involve causation. Others are false, but we should keep them anyway and pretend as if they are true, because doing so is useful. Pretending that our time-involving agentive thoughts are true is an efficient way of codifying causal structure that is relevant to agency. This version of agentive fictionalism is thus committed to a degree of temporal fictionalism. That's because it involves continuing to use false, but useful temporal thoughts, namely those that are involved in certain agentive thoughts. Note that there is no problem with entering the temporal fiction in this case. That's because there are some agentive thoughts that are strictly true. To get into the temporal fiction, then, one must simply token a true agentive thought involving causal structure in the first instance.

Indeed, the picture of agency sketched here affords us with a way to revive temporal fictionalism more generally. As discussed in the previous chapter, the central problem for temporal fictionalism was that it seemed to require agency. For, recall, in order to be able to make sense of how we could choose to engage in the temporal fiction, we need to first make sense of our having choices at all. Vindicating agentive realism presents a way to overcome the problem. Moreover, it motivates temporal fictionalism over the eliminativist alternative. Indeed, we think, it motivates temporal fictionalism over the other alternatives—temporal noncognitivism and temporal realism—that we outlined in Chapter Nine. For it is a view on which we need make no revolutionary claims about the content of our temporal thoughts, or what makes them true. We admit that such thoughts are false. But this does not have any disastrous consequences. There is, we submit, no real downside to having false temporal thoughts.

Why can't the strategy we have used to recover agency be used to recover time as well? Well, what would the analogous strategy be? Broadly, there are two options. First, one might argue that although our temporal thoughts are false, there are nearby thoughts that are strictly speaking true and that can play the same role for us in our everyday lives. Second, one might argue that our concepts of time are in fact satisfied by some underlying structure that exists in the world (in the way that we are imagining for causation and deliberative structure).

The second option relates to the idea, discussed in Chapter Seven, that approximate spacetime structure might be sufficient to satisfy everyday notions of time after all, even if that structure does not feature a C-series. We have nothing new to add to this option beyond what we have already said and so we set it aside. We have some reservations about the first option. In particular, it is unclear what we might trade temporal thoughts in for that could play the same role in our lives. The only option seems to be causal thoughts but, as discussed, we have reason not to trade temporal thoughts for causal thoughts, given that the former provide a useful (if false) way of summarising causal information.

Ultimately, though, we are open to the idea that the strategy we supply here might be used for temporal thoughts as well. For that would be just another way of achieving the goal of Part Three. For it would make possible that time does not exist (because the concept is not satisfied) and yet all our everyday practices built up around time can be preserved without temporal thought. This is really a version of the eliminativist strategy discussed in the previous chapter (as noted we have some basis for preferring the fictionalist option).

11.6. Temporal Recovery

In response to the conceptual argument we maintained that the supportive structure for agency need not be thought of in terms of a C-series, but that it can be thought of in causal terms instead. In this way, to a certain extent we end up breaking the connection between agency and the folk notion of time. Instead, what we are proposing is that the folk notion of agency is connected to notions of deliberation, but it is left open whether deliberation is understood in temporal or causal terms. If time exists, then supportive structure for deliberation can take a temporal form. But in the absence of time, the supportive structure for deliberation can take a causal form.

We recognise that, in saying this, we are making an empirical bet about the concept of agency. We are betting that the folk concept of agency is, in the end, not an essentially temporal concept, it is either a causal concept or, at worst, something like a disjunctive or conditional concept, which gives the notion flexibility in the face of discoveries about the world, relating in particular to time. In Chapter Two, however, we considered some similar empirical bets about the folk concept of time. For instance, we considered the option that people would continue to say that time exists even if they report in an experimental setting that they wouldn't make such judgements. By making one bet and setting aside another, aren't we guilty of holding a double standard?

There are two things to say here. First, there are bets and there are bets. The bet we are making is made without any evidence of what people would, or would not say, about agency, when presented with particular scenarios involving time, on the one hand, and causation on the other. By contrast, the bets that we considered in Chapter Two are bets made in the face of evidence that is contrary to the bet in question. One must bet that, despite what people report they would say (which is some evidence of what they would say), people would in fact do something different. We find our bet to be less epistemically risky than the bets we set aside in Chapter Two. It is one thing to bet, without evidence that the world will be thus and so. It is quite another thing to bet against the evidence.

Ultimately, then, we think our bet is better than the ones we considered in Chapter Two. As such, we are more inclined to make the kind of bet we make on agency, than the associated bets about time. This is not to say that one cannot make the other kind of bet. It is only to point out a betting asymmetry. The asymmetry explains why we think that the strategy we use for securing agency in this chapter does not straightforwardly carry over to time, since the analogous strategy for time would require one to bet against the evidence outlined in Part One of this book.

Second, it is possible to understand what we do in this chapter in a manner that does not require taking a bet on what the folk concept of agency might be. Set aside what the folk think. It is possible to view everything we say about agency as an exercise in conceptual engineering. What we show is how to *engineer* a concept of agency that can be satisfied in a world that lacks time: namely the concept stated above. Thus, agency—in the sense described— remains, even if time, in the folk sense, does not.

One might balk at this. Why should we take the new concept—the one we have engineered—to be a concept of agency at all? Assuming that our understanding of agency is, in the first instance, given to us by our folk concept of the same, why think that a re-engineered concept of agency is worthy of the name? Presumably, it is because the engineered concept bears some measure of continuity with the folk concept, one might think. But then, even a project of conceptual engineering requires some empirical understanding of our folk concepts in order to make the inevitable case for continuity. If so, then one might worry that we cannot avoid the need for further empirical investigation of our folk concepts of agency quite so easily.

Ultimately, we don't care whether the concept outlined above should be *called* agency or not, and thus it doesn't matter to us whether the concept is sufficiently similar to the folk notion. The important thing is that the notion we have outlined above—call it schmagency if you must—seems capable of doing the right kind of work for beings like us. In particular, it seems to be the kind of thing that can support decision-making, and that can support our self-conception as beings who can freely choose to intervene in the world, and thereby intervene as a result of that choice. Since this is what ultimately matters, rather than the continuity of any particular concept, we take our conceptual engineering to be good enough to blunt the implications of the discovery that there is no C-series.

Why can't a similar strategy of conceptual engineering be used to recover time? Perhaps it can. Note, though, that causation won't obviously do as the basis for time, since it is not at all obvious that we can conceptually engineer a concept of time around the notion of causation. Indeed, to a certain extent a project in this neighbourhood has been tried and found wanting (we think, here, of the causal theories of time, which were rejected, in part, because it is difficult to understand temporal concepts in causal terms). A better bet would be to appeal to the notion of approximate spacetime discussed in Chapter Seven.

Ultimately, though, we don't need to show that time *cannot* be conceptually re-engineered in such a manner that the concept can be satisfied even if we make discoveries of the kind that might be on the horizon in physics. All we need is the weaker claim that one can conceptually reengineer agency, without also doing the same to time. Since if that's true, then there can be agency without time, which is the result that we ultimately seek.

11.7. Conclusion

In sum, then, what we end up with is a picture on which agency is real and grounded in causation. Some agentive thoughts are false because they explicitly token temporal thoughts. We should behave as if the temporal thoughts are true, however, and so we should continue to behave as though those agentive thoughts are true, which token those temporal thoughts. We should do so just when those agentive thoughts would, literally be true if instead they tokened some appropriate causal thought. Our picture of agency has a limited degree of fictionalism, despite generally being realist in nature. But a limited degree of fictionalism is acceptable, so long as there is enough agency to underwrite a genuine choice to engage in the temporal fiction. We conclude, then, that the existence of agency is not at odds with discoveries that would lead people to reject the existence of time. One can continue being an agent, and continue believing in agency, and acting as an agent, even without time, so long as causation exists.

This concludes our argument in favour of the claim that time might not exist. Our argument targeted three main reasons why one might think that time must exist. First, that folk concepts of time are immune to error. Second, that the discoveries that would lead one to reject the existence of time are incompatible with what we know about the world from science and, third, that the same discoveries are at odds with the existence of agency. By defeating all three reasons, we are now able to draw our main conclusion: that time, in the everyday sense, the sense we care about, might not exist. Of course, it might exist; but there is an epistemic crack around time into which a timeless picture of reality may be inserted.

We have also shown that the discovery that time does not exist need not be particularly startling. For, in the end, agency is perfectly safe even in a world without time, so long as causation remains. In a sense, it is business as usual: we can continue to speak, think and act just as we ever did. Indeed, once we see this, a purely philosophical argument against the existence of time starts to emerge. Time in the folk sense starts to look like a metaphysical dangler. While it may be useful, in an everyday sense, to have temporal thoughts because they ease the cognitive load of laying out causal structure, nothing is gained from believing that the thoughts track some real feature of reality. Time may be an extravagance that we can well afford to do without. Whether this purely philosophical argument succeeds, however, is a tale for another time.

Future Directions

In this book, we have argued for the following three main claims:

[1] Folk concepts of time are not immune to error.
[2] Discoveries that would lead people to reject the existence of time are compatible with what we know about the world from science.
[3] Discoveries that would lead people to reject the existence of time are compatible with agency.

These three claims lead us to our conclusion: time might not exist. Time might not exist because the reasons for thinking that this is an epistemically impossibility don't ultimately hold up. Whether time exists, and whether the absence of time is a metaphysical or physical possibility is not something we have argued for, though proper parts of the authorship are sympathetic to one or more of these claims.

Along the way, we have argued for a number of secondary claims:

[1] There is no single folk conception of time
[2] We have reason to doubt that spacetime is an emergent phenomenon because we lack a viable metaphysics of spacetime emergence.
[3] There can be causation without a C-series.
[4] In the absence of a C-series, the combination of agentive realism and temporal fictionalism is the best approach to temporal and agentive thought and talk.
[5] Even if time does not exist, this need make no difference to our everyday lives.

In no sense do we take our book to be the final word on timelessness. It is, rather, a preliminary foray into the question of how we should make sense of the world if our ordinary notions of time are shown to be in error. There is much work to be done. In closing, then, we will flag three areas for future work.

Out of Time: A Philosophical Study of Timelessness. Sam Baron, Kristie Miller, and Jonathan Tallant, Oxford University Press.
© Sam Baron, Kristie Miller, and Jonathan Tallant 2022. DOI: 10.1093/oso/9780192864888.003.0012

First, there is a great deal more to be said about the metaphysics of space-time emergence. We have argued for a particular view, but ultimately we recognise that the question of whether and how spacetime emerges will outrun the arguments offered in this book. We have not, for instance, provided a detailed assessment of the various approaches to QG. Nor have we considered the metaphysical implications of every approach to QG for our everyday notions of time and causation. There is a great deal of work to be done in exploring the metaphysical implications of this growing area of physics. We anticipate that a central component of this work will involve the development of new metaphysical models of the world. We welcome and look forward to new metaphysical work in this area.

Second, the relationship between causation, time and counterfactual dependence demands a much more broader and deeper discussion than we have afforded it here. We have gone some way toward answering the question of whether causation requires time, but there is much more to say. In a certain sense, we expect this issue to dovetail with the last: as the physics of QG develops we expect causation, and not just time, to be of central importance. New models of counterfactual dependence are also likely to be required, as the picture of the world potentially shifts from one that features fundamental spacetime to one that does not.

Third, we have not said much about the psychology of time, despite this being a burgeoning area of philosophical interest. There are, however, important and obvious connections between our discussion of observation and confirmation in the absence of time, and the perception and experience of time. Indeed, we suspect that the relationship between time and mind runs deeper still. The question of how the world is fundamentally structured, and whether time in some sense is a part of that structure, may well cast new light on the nature of consciousness. The contribution that the metaphysics of time makes to the philosophy of mind remains an underexplored topic, but one that the physics and metaphysics of QG may well bring to the fore.

Bibliography

Adleberg, T., Thompson, M., and Nahmias, E. (2015). Do Men and Women Have Different Philosophical Intuitions? Further Data. *Philosophical Psychology*, 28(5): 615–41. doi:10.1080/09515089.2013.878834.

Albert, D. Z. (2000). *Time and Chance*. Cambridge, Mass: Harvard University Press.

Ambjørn, J., Jurkiewicz, J. and Loll, R. (2001). Dynamically Triangulating Lorentzian Quantum Gravity. *Nuclear Physics B* 610(1–2): 347–82.

Ambjørn, J., Jurkiewicz, J. and Loll, R. (2004). Emergence of a 4D World from Causal Quantum Gravity. *Physical Review Letters* 93: 131301.

Anderson, E. (2006). Relational Particle Models: 1. Reconciliation with Standard Classical and Quantum Theory. *Classical and Quantum Gravity*, 23(7): 2469–90.

Anderson, E. (2009). Records Theory. *International Journal of Modern Physics D*, 18(4): 635–67.

Anderson, E. (2012a). Problem of Time in Quantum Gravity. *Annalen der Physik*, 524(12): 757–86.

Anderson, E. (2012b). The Problem of Time in Quantum Gravity. In V. R. Frignanni (ed.), *Classical and Quantum Gravity: Theory, Analysis and Applications* (pp. 1–25). New York: Nova.

Armour-Garb, B. and Woodbridge, J. A. (2015). *Pretense and Pathology: Philosophical Fictionalism and Its Applications*. CUP.

Bain, J. (2013). The Emergence of Spacetime in Condensed Matter Approaches to Quantum Gravity. *Studies in History and Philosophy of Modern Physics*, 44: 338–45.

Barbour, Julian. (1999). *The End of Time*. (Oxford: Oxford University Press).

Barbour, J. (1994a). The Timelessness of Quantum Gravity: I. The Evidence from the Classical Theory. *Classical Quantum Gravity*, 11(12): 2853–73.

Barbour, J. (1994b). The Timelessness of Quantum Gravity: II. The Appearance of Dynamics in Static Configurations. *Classical Quantum Gravity*, 11(12): 2875–97.

Bardon, A. (2013). *A Brief History of the Philosophy of Time* (Oxford: Oxford University Press).

Baron, S. (2013). Talking about the Past, *Erkenntnis*, 78(3): 547–560.

Baron, S. (2019). The Curious Case of Spacetime Emergence. *Philosophical Studies*, 177: 2207–26.

Baron, S. (2021). Parts of Spacetime. *American Philosophical Quarterly* 58(4).

Baron, S. Chua, R., Miller, K., and Norton, J. (2019). Much ado about aboutness. *Inquiry*, https://doi.org/10.1080/0020174X.2019.1592705.

Baron, S., Colyvan, M., and Ripley, R. (2017). How Mathematics Can Make a Difference. *Philosophers' Imprint*, 17(3): 1–29.

Baron, S., Colyvan, M., and Ripley, R. (2020). A Counterfactual Approach to Explanation in Mathematics. *Philosophia Mathematica*, 28(1): 1–34.

Baron, S., Cusbert, J., Farr, M., Kon, M., and Miller, K. (2015). Temporal Experience, Temporal Passage and the Cognitive Sciences. *Philosophy Compass*, x (8): 56–571.

Baron, S. and Miller, K. (2015a). What Is Temporal Error Theory? *Philosophical Studies*, 172 (9): 2427–44.

Baron, S. and Miller, K. (2015b). Our Concept of Time. In *Philosophy and Psychology of Time* edited by B. Mölder, V. Arstila, and P. Ohrstrom. Springer, pp. 29–52.

Baron, S., Evans, P., and Miller, K. (2010). From Timeless Physical Theories to Timelessness. *Humana Menta*, 13: 32–60.

Baron, S., Miller., K., and Tallant, J. (2021). Temporal Fictionalism for a Timeless World. *Philosophy and Phenomenological Research*, 102(2): 281–301.

Barrett, J. A. (1999). The Quantum Mechanics of Minds and Worlds. New York: Oxford University Press.

Barnes, E. (2012). Emergence and Fundamentality, *Mind* 121(484): 873–901.

Bedau, Mark (2008). Is Weak Emergence Just in the Mind? *Minds & Machines* 18: 443–59.

Benedetti, D., Machado, P. F., and Saueressig, F. (2009). Asymptotic Safety in Higher-Derivative Gravity. *Modern Physics Letters A*, 24(28): 2233–41.

Bigelow, John (1996). Presentism and Properties, *Philosophical Perspectives*, 10(1): 35–52.

Block, N. (1986). Advertisement for a Semantics for Psychology. *Midwest Studies in Philosophy*, 10: 615–78.

Boroditsky, L. (2001). Does Language Shape Thought? English And Mandarin Speakers' Conceptions of Time. *Cognitive Psychology*, 43: 1–22.

Boroditsky, L., Fuhrman, O., and McCormick, K. (2011). Do English and Mandarin Speakers Think about Time Differently? *Cognition*, 118, 123–9.

Bourne, C. (2006). *A Future for Presentism*. Oxford: Oxford University Press.

Braddon-Mitchell, D. (2013). Against the Illusion Theory of Temporal Phenomenology. *CAPE Studies in Applied Ethics*, II: 211–33.

Braddon-Mitchell, D. and Miller, K. (2019). Quantum Gravity, Timelessness, and the Contents of Thought. *Philosophical Studies*, 176(07): 1807–29.

Braddon-Mitchell, D. and Miller, K. (2006). The Physics of Extended Simples. *Analysis*, 66: 222–6.

Braddon-Mitchell, D. and Nola, R. (2008). *Conceptual Analysis and Philosophical Naturalism*. Bradford Press.

Brenner, A. (2015). Mereological Nihilism and Theoretical Unification. *Analytic Philosophy*, 56(4): 318–37.

Broad, C. D. (1923). *Scientific Thought* (London Routledge & Kegal Paul).

Buckwalter, W. and Stich, S. (2014). Gender and Philosophical Intuition. In *Experimental Philosophy*, Volume 2, edited by Joshua Knobe and Shaun Nichols, New York: Oxford University Press, pp. 307–46. doi:10.1093/acprof:osobl/9780199927418.003.0013.

Butterfield, J. and Isham, C. (1999). On the Emergence of Time in Quantum Gravity. In *The Arguments of Time*. Jeremy Butterfield (ed.), 111–68 (Oxford; New York: Oxford University Press).

Button, T. (2007). Every Now and Then, No-Futurism Faces No Sceptical Problems. *Analysis*, 67(4): 325–32.

Callender, C. (2008). The Common Now. Philosophical *Issues*, 18(1): 339–61.

Callender, C. (2017). *What Makes Time Special?* Oxford University Press.

Callender, C. and Suhler, C. (2012). Thank Goodness That Argument Is Over: Explaining the Temporal Value Asymmetry. *Philosophers' Imprint*, 12: 1–16.

Cameron, R. P. (2011). *Truthmaking for Presentists, Oxford Studies in Metaphysics*, Vol. 6, ed. Karen Bennett and Dean Zimmerman, Oxford: Oxford University Press: 55–100.

Cameron, R. (2008). How to Be a Truthmaker Maximalist. *Noûs*, 42(3): 410–21.

Cameron, R. P. (2015). *The Moving Spotlight: An Essay on Time and Ontology*. Oxford University Press.

Caplan, B. and Sanson, D. (2010). The Way Things Were. *Philosophy and Phenomenological Research*, 81(1): 24–39.

Carroll, Sean (2003). *Spacetime and Geometry: An Introduction to General Relativity* (UK: A&S Academic Science).

Cartan, Élie (1923). Sur les variétés à connexion affine et la théorie de la relativité généralisée (Première partie). *Annales Scientifiques de l'École Normale Supérieure*, 40: 325.

Casasanto, D. and Bottini, R. (2014). Mirror Reading Can Reverse the Flow of Time. *Journal of Experimental Psychology: General*, 143: 473–9.

Casati, R. and Varzi, A. (1999). *Parts and Places: The Structure of Spatial Representation* (Cambridge, Massachusetts: MIT Press).

Chalmers, D. (2004). Epistemic Two Dimensional Semantics, *Philosophical Studies*, 118: 153–226.

Chalmers, D. J. and Jackson, F. (2001). Conceptual Analysis and Reductive Explanation. *Philosophical Review* 110(3): 315–61.

Chalmers, D. (2008). Strong and Weak Emergence. In *The Re-Emergence of Emergence: The Emergentist Hypothesis from Science to Religion*, Philip Clayton and Paul Davies (eds.) (Oxford: Oxford University Press).

Chalmers, D. (2012). *Constructing the World*. Oxford University Press.

Chalmers, David (2021). Finding Space in a Nonspatial World. In *Philosophy Beyond Spacetime* (Baptiste Le Bihan and Nick Huggett, eds), Oxford University Press.

Chalmers, David J. (2004b). The Representational Character of Experience. In Brian Leiter (ed.), *The Future for Philosophy*, pp. 153–81 (Oxford: Oxford University Press).

Chrisman, M. (2007). A Dilemma for Moral Fictionalism. *Philosophical Books* 49(1): 4–13.

Contessa, G. (2014). One's a Crowd: Mereological Nihilism Without Ordinary-Object Eliminativism. *Analytic Philosophy* 54(4): 199–221.

Craig, W. L. (2000). *The Tensed Theory of Time: A Critical Examination* (Dordrecht: Kluwer Academic Publishers).

Crane, Tim (2013). *The Objects of Thought*. Oxford: Oxford University Press.

Crisp, T. (2003). Presentism. In Michael J. Loux and Dean W. Zimmerman (eds.), *The Oxford Handbook of Metaphysics*. Oxford University Press.

Crisp, Thomas (2007). Presentism and the Grounding Objection, Noûs 41/1: 90–109.

Crowther, Karen (2016). *Effective Spacetime: Understanding Emergence in Effective Field Theory and Quantum Gravity* (Springer: Heidelberg).

Crowther, Karen (2018). Inter-Theory Relations in Quantum Gravity: Correspondence, Reduction and Emergence. *Studies in History and Philosophy of Modern Physics* 63: 74–85.

Cummins, R. (1989). *Meaning and Mental Representation* (Cambridge: Bradford Books/MIT Press).

Cusbert, J. and Miller, K. (2018). The Unique Groundability of Temporal Facts. *Philosophy and Phenomenological Research*, 97(20): 410–32.

Dainton, B. (2000). *Stream of Consciousness: Unity and Continuity in Conscious Experience* (Routledge).

Dainton, B. (2008). *The Phenomenal Self*. Oxford University Press.

Dautcourt, G. (1964). Die Newtonische Gravitationstheorie als strenger Grenzfall der allgemeinen Relativitätstheorie. *Acta Physica Polonica*, 65: 637–46.

Davies, P. (1995). *About Time: Einstein's Unfinished Revolution* (Harmondsworth: Penguin).

Deasy, D. (2015). The Moving Spotlight Theory. *Philosophical Studies*, 172(8): 2073–89.

Deasy, D. (2017). What is Presentism. *Noûs* 51(2): 378–97.

Deng, N. (2013). On Explaining Why Time Seems to Pass. *Southern Journal of Philosophy*, 51(3): 367–82.

Deng, N. (2018). What is Temporal Ontology? *Philosophical Studies* 175(3): 793–807.

DeWitt, Bryce S. (1967). Quantum Theory of Gravity. I. The Canonical Theory. *Physical Review*, 160(5): 1113–48.

Dittrich, Bianca and Thiemann, Thomas (2009). Are the Spectra of Geometrical Operators in Loop Quantum Gravity Really Discrete? *Journal of Mathematical Physics* 50, 102503.

Dorr, C. (2002). The Simplicity of Everything. PhD thesis, Princeton University.

Dowe, P. (1992). Wesley Salmon's Process Theory of Causality and the Conserved Quantity Theory. *Philosophy of Science* 59(2): 195–216.

Dowker, Faye (2006). Causal Sets as Discrete Spacetime. *Contemporary Physics* 47: 1–9.

Dowker, Faye (2014). The Birth of Spacetime Atoms as the Passage of Time. *Annals of the New York Academy of Sciences* 1326: 18–25.

Dretske, F. (1981). *Knowledge and the Flow of Information*. Harvard: MIT Press.

Dretske, F. (1983). Precis of Knowledge and the Flow of Information. *Behavioral and Brain Sciences*, 6: 55–63.

Eagleman, D. (2008). Human Time Perception and Its Illusions. *Current Opinion in Neurobiology*, 18: 131–6.

Eagleman, D. M. (2010). The Strange Mapping Between the Timing of Neural Signals And Perception. In Issues of Space and Time in Perception and Action, R. Nijhawan (ed). Cambridge University Press.

Eagleman, D. M. (2011). *Incognito: The Hidden Life of the Brain* (Canongate Books).

Earman, John (1989). *World Enough and Space-Time: Absolute versus Relational Theories of Space and Time* (Cambridge, MA: MIT Press).

Earman, J. (2002). Thoroughly Modern Mctaggart: Or, What Mctaggart Would Have Said If He Had Read the General Theory of Relativity. *Philosopher's Imprint*, 2(3): 1–28.

Ehlers J. (1973). Survey of General Relativity Theory. In: Israel W. (ed.), Relativity, Astrophysics and Cosmology. Astrophysics and Space Science Library (A Series of Books on the Recent Developments of Space Science and of General Geophysics and Astrophysics Published in Connection with the Journal Space Science Reviews), vol 38. Springer, Dordrecht.

Ehlers, Jürgen (1997). Examples of Newtonian Limits of Relativistic Spacetimes. *Classical Quantum Gravity* 13: A119–A26.

Fair, D. (1979). Causation and the Flow of Energy. *Erkenntnis*, 14: 219–50.

Farkas, Katalin (2008). Phenomenal Intentionality Without Compromise. *The Monist*, 91 (2): 273–93.

Farr, M. (2012). On A- and B-Theoretic Elements of Branching Spacetimes. *Synthese*, 188 (1): 85–116.

Field, H. (1977). Logic, Meaning and Conceptual Role. *Journal of Philosophy*, 69: 379–408.

Fodor, J. (1987). *Psychosemantics: The Problem of Meaning in the Philosophy of Mind* (Cambridge, MA: MIT/Bradford).

Fodor, J. (1990). *A Theory of Content and Other Essays* (Cambridge, MA: MIT/Bradford Press).

Forbes, G. (2016). The Growing Block's Past Problems. *Philosophical Studies* 173(3): 699–709.

Forbes, G. and Briggs, R. (2017). The Growing-Block: Just One Thing After Another? *Philosophical Studies*, 174(4): 927–43.

Forrest, P. (2006). Uniform Grounding of Truth and the Growing Block Theory: A Reply to Heathwood. *Analysis*, 66(2): 161–3.

Friesdorf, Rebecca, Conway, Paul, and Gawronski, Bertram (2015). Gender Differences in Responses to Moral Dilemmas: a Process Dissociation Analysis. *Personality and Social Psychology Bulletin*, 41(5): 696–713. doi:10.1177/0146167215575731.

Fuhrman, O. and Boroditsky, L. (2010). Cross-Cultural Differences in Mental Representations of Time: Evidence from an Implicit Non-Linguistic Task. *Cognitive Science*, 34: 1430–145.

Gale, R. M. (1968). *The Language of Time* (Virginia: Humanities Press).

Gallagher, S. (2003). Sync-ing in the Stream of Experience: Time-Consciousness in Broad, Husserl, and Dainton. *PSYCHE: An Interdisciplinary Journal of Research On Consciousness* 9.

Gevers, W., Reynvoet, B., and Fias, W. (2003). The Mental Representation of Ordinal Sequences Is Spatially Organised. *Cognition*, 87(3): B87–95.

Gödel, Kurt (1949). An Example of a New Type of Cosmological Solutions of Einstein's Field Equations of Gravitation. *Reviews of Modern Physics*, 21(3): 447–50.

Harman, G. (1982). Conceptual Role Semantics. *Notre Dame Journal of Formal Logic*, 23: 242–57.

Harman, G. (1990). The Intrinsic Quality of Experience. *Philosophical Perspectives*, 4: 31–52.

Havas, P. (1964). Four-Dimensional Formulations of Newtonian Mechanics and Their Relation to the Special and General Theory of Relativity. *Reviews of Modern Physics*, 36 (4): 938–65.

Healey, R. (2002). Can Physics Coherently Deny the Reality of Time? In C. Callender (ed.), *Time, Reality & Experience*, pp. 293–316 (Cambridge: Cambridge University Press).

Hoerl, C. (2014). Do We (Seem to) Perceive Passage? *Philosophical Explorations*, 17: 188–202.

Hohwy, J., Paton B., and Palmer, C. (2016). Distrusting the Present. *Phenomenology and the Cognitive Sciences* 15(3): 315–35.

Holcombe, A. O. (2015). The Temporal Organisation of Perception. In *The Oxford Handbook of Perceptual Organisation*. Edited by J Wagemans (Oxford University Press).

Horgan, Terence E., Tienson, John L., and Graham, George (2004). Phenomenal Intentionality and the Brain in a Vat. In Richard Schantz (ed.), *The Externalist Challenge* (De Gruyter).

Horwich, P. (2005). *Reflections on Meaning* (Oxford: Oxford University Press).

Hu, Bei-Lok (2009). Emergent/Quantum Gravity: Macro/Micro Structures of Spacetime. *Journal of Physics: Conference Series* 174(1): 1–16.

Huemer, M. and Kobitz, B. (2003). Causation as Simultaneous and Continuous. *Philosophical Quarterly* 53(213): 556–65.

Huggett, Nick (2017). Target Space ≠ Space. *Studies in History and Philosophy of Modern Physics*, 59: 81–8.

Huggett, Nick and Wüthrich, Christian (2013). Emergent Spacetime and Empirical (in) coherence. *Studies in History and Philosophy of Modern Physics*, 44: 276–85.

Humphreys, Paul (2016). *Emergence: A Philosophical Account* (Oxford: Oxford University Press).

Ingram, D. (2016). The Virtues of Thisness Presentism. *Philosophical Studies*, 173(11): 2867–88.

Ingram, D. (2019). *Thisness Presentism: An Essay on Time, Truth, and Ontology* (Routledge).

Ingthorsson, R. (2016). *McTaggart's Paradox* (Routledge).

Ismael, J. (2012). Decision and the Open Future. In *The Future of the Philosophy of Time*, edited by A Bardon, Routledge: 149–69.

Jackson, F. (2004). Why We Need A-Intensions. *Philosophical Studies*, 118: 257–77.

Jaszczolt, K. M. (2013). Temporality and Epistemic Commitment: An Unresolved Question. In K. M. Jaszczolt and L. de Saussure (eds), *Time: Language, Cognition, and Reality*, pp.193–209 (Oxford: Oxford University Press).

Johnson, M. W. and Bickel, W. K. (2002). Within-Subject Comparison of Real And Hypothetical Money Rewards in Delay Discounting. *Journal of the Experimental Analysis of Behavior*, 77: 129–46.

Joyce, R. (2011). Moral Fictionalism. *Philosophy Now*, 82: 14–7.

Kalderon, M. E. (2005). *Moral Fictionalism* (Clarendon Press).

Keller, Simon (2004). Presentism and Truthmaking, *Oxford Studies in Metaphysics, Vol. 1*, ed. Dean W. Zimmerman, pp. 83–104 (Oxford: Oxford University Press).

Kim, Minsun, and Yuan Yuan (2015). No Cross-Cultural Differences in the Gettier Car Case Intuition: A Replication Study of Weinberg et al. 2001. *Episteme*, 12(03): 355–61. doi:10.1017/epi.2015.17.

Knox, Eleanor (2011). Newton-Cartan Theory and Teleparallel Gravity: The Force of a Formulation. *Studies in the History and Philosophy of Modern Physics*, 42: 264–75.

Knox, Eleanor (2014). Newtonian Spacetime Structure in Light of the Equivalence Principle. *British Journal for the Philosophy of Science*, 65(4): 863–80.

Knox, E. (2018). Physical Relativity From A Functionalist Perspective. *Studies in History and Philosophy of Modern Physics* 67: 118–124. https://doi.org/10.1016/j.shpsb.2017.09.008.

Koslicki, K. (2015). The Coarse-Grainedness of Grounding, in K. Bennett and D. Zimmerman (eds), *Oxford Studies in Metaphysics Volume 9*, pp. 306–44 (Oxford: Oxford University Press).

Kriegel, Uriah (2011). *The Sources of Intentionality* (Oxford: Oxford University Press).

Kriegel, Uriah (2013). The Phenomenal Intentionality Research Program. In U. Kriegel (ed.), *Phenomenal Intentionality* (Oxford: Oxford University Press).

Kuchar, Karel V. (1992). Canonical Quantum Gravity. In R. J. Gleiser, C. N. Kozameg, and O. M. Moreschi (eds), *General Relativity and Gravitation 1992: Proceedings of the Thirteenth International Conference on General Relativity and Gravitation*, 119–50 (Bristol and Philadelphia: Institute of Physics Publishing).

Kuehni, Rolf G. and Hardin, C.L. (2010). Churchland's Metamers. *British Journal for the Philosophy of Science*, 61(1): 81–92. doi:10.1093/bjps/axp021.

Künzle, H. (1976). Covariant Newtonian Limits of Lorentz Space-Times. *General Relativity and Gravitation*, 7(5): 445–57.

Kushnir, T., Gopnik, A., Lucas, C., and Schulz, L. (2010). Inferring Hidden Causal Structure. *Cognitive Science*, 34(1): 148–60.

Lagnado, D. A. and Sloman, S. A. (2004). The Advantage of Timely Intervention. *Journal of Experimental Psychology: Learning, Memory, and Cognition*, 30: 856–76.

Lagorio, C. H. and Madden, G. J. (2005). Delay Discounting of Real and Hypothetical Rewards. *Behav Processes*, 69(2): 173–87.

Lai, V. T. and Boroditsky, L. (2013). The Immediate and Chronic Influence of Spatio-Temporal Metaphors on the Mental Representation of Time in English, Mandarin, and Mandarin-English Speakers. *Frontiers in Psychology*, 4: 142. doi: 10.3389/fpsyg.2013.00142.

Lam, V. and Esfeld, M. (2013). A Dilemma for the Emergence of Spacetime in Canonical Quantum Gravity. *Studies in History and Philosophy of Modern Physics*, 44: 286–93.

Lam, V. and Wüthrich, C. (2018). Spacetime is as Spacetime Does. *Studies in History and Philosophy of Modern Physics* 64: 39–51.

Land, Edwin H. (1983). Recent Advances in Retinex Theory and Some Implications for Cortical Computations: Color Vision and the Natural Image. *Proceedings of the National Academy of Sciences*, 80(16): 5163–9.

Latham, A. J. (2019). Indirect Compatibilism. PhD Dissertation. https://philpapers.org/rec/LATIC.

Latham, A. J. and Miller, K. (2020a). Quantum Gravity, Timelessness and the Folk Concept of Time. *Synthese*, 198(10): 9453–78. https://doi.org/10.1007/s11229-020-02650-y.

Latham, A. J. and Miller, K. (2020b). Time in a One Instant World. *Ratio*, 33(3): 145–54. DOI:10.1111/rati.12271.

Latham, A. J. and Miller, K. (2021). Are the Folk Functionalists about Time? *Southern Journal of Philosophy*. http://doi.org/10.1111/sjp.12441.

Latham, A. J., Miller, K., and Norton, J. (2019). Is Our Naïve Theory of Time Dynamical? *Synthese*, 198(5): 4251–71. DOI: 10.1007/s11229-019-02340-4[2]

Latham, A. J., Miller, K., and Norton, J. (2020a). An Empirical Investigation of the Role of Direction in Our Concept of Time. *Acta Analytica*, 36(1): 25–47. DOI: 10.1007/s12136-020-00435-z

Latham, A. J., Miller, K., and Norton, J. (2020b). Do the Folk Represent Time as Essentially Dynamical? *Inquiry*. https://doi.org/10.1080/0020174X.2020.1827027

Latham, A. J., Miller, K., and Norton, J. (2020c). An Empirical Investigation of Purported Passage Phenomenology. *The Journal of Philosophy*, 117(7): 353–86.

Le Bihan, Baptiste (2018a). Spacetime Emergence in Contemporary Physics. *Disputatio*, 49 (10): 71–85.

Le Bihan, Baptiste (2018b). Priority Monism Beyond Spacetime. *Metaphysica*, 19(1): 95–111.

Le Bihan, Baptiste and Read, James (2018). Duality and Ontology. *Philosophy Compass*, 13 (12): e12555.

Le Bihan, Baptiste and Lineman, Niels (2019). Have We Lost Spacetime on the Way? Narrowing the Gap Between General Relativity and Quantum Gravity. *Studies in History and Philosophy of Modern Physics*, 65: 112–21.

Lee, G. (2014). Temporal Experience and the Temporal Structure of Experience. *Philosophers' Imprint*.

Leininger, L. (2021). 'Temporal B-Coming: Passage Without Presentness.' *Australasian Journal of Philosophy*, 99(1): 130–47.

Le Poidevin, R. (1991). *Change, Cause and Contradiction: A Defence of the Tenseless Theory of Time* (London: Macmillan Press Ltd).

Le Poidevin, R. (2007). *The Images of Time: An Essay on Temporal Representation* (Oxford: Oxford University Press).

Lepore, E. (1994). Conceptual Role Semantics. In *A Companion to the Philosophy of Mind*, ed. S. Guttenplan (Oxford: Blackwell).

Lewis, David (1970). How to Define Theoretical Terms. *Journal of Philosophy*, 67: 427–46.

Lewis, David (1972). Psychophysical and Theoretical Identifications. *Australasian Journal of Philosophy*, 50: 249–58.

Lewis, David (1973). Causation. *Journal of Philosophy*, 70(17): 556–67.

Lewis, David (1979). Counterfactual Dependence and Time's Arrow. *Noûs*, 13(4): 455–76.

Lewis, David (1986). *Philosophical Papers, Volume II* (Oxford University Press).

Liggins, D. (2008). Nihilism without Self-Contradiction. In *Royal Institute of Philosophy Supplement* 62(62). Cambridge: Cambridge University Press, pp. 177–96.

Loar, Brian (2003). Qualia, Properties, Modality. *Philosophical Issues*, 13(1): 113–29.

Loewer, B. (2012). Two Accounts of Laws and Time. *Philosophical Studies*, 160(1): 115–37.

Ludlow, P. (1999). *Semantics, Tense, and Time* (MIT Press).

Machery, E., Stich, S., Rose, D., Chatterjee, A., Karasawa, K., Struchiner, N., Sirker, S., Usui, N., and Hashimoto, T. (2015). Gettier Across Cultures, *Noûs*, 51(3): 645–64. doi:10.1111/nous.12110

Machery, E., Mallon, R., Nichols, S., and Stich, S. P. (2004). Semantics, Cross-Cultural Style, *Cognition*, 92(3): B1–B12. doi:10.1016/j.cognition.2003.10.003

Malament, David B. (1977). The Class of Continuous Timelike Curves Determines the Topology of Spacetime. *Journal of Mathematical Physics*, 18: 1399.

Maloney, J. (1994). Content: Covariation, Control and Contingency. *Synthese*, 100: 241–90.

Markosian, N. (2004). A Defense of Presentism. In D. Zimmerman (ed.) *Oxford Studies in Metaphysics* Volume I (Oxford: Oxford University Press).

Markosian, N. (2014). A Spatial Approach to Mereology. In S. Kleinshmidt (ed.), *Mereology and Location* (Oxford: Oxford University Press).

Maudlin, T. (2002). Thoroughly Muddled McTaggart or How to Abuse Gauge Freedom to Generate Metaphysical Monstrosities. *Philosopher's Imprint*, 2(4): 1–23.

Maudlin, T. (2007). *The Metaphysics Within Physics* (Oxford: Clarendon Press).

McTaggart, J. M. E. (1908). The Unreality of Time. *Mind*, 17(68): 457–74.

Mellor, D. H. (1998). *Real time II* (London: Routledge).

Mendelovici, Angela (2010). *Mental Representation and Closely Conflated Topics*. Dissertation, Princeton University.

Merricks, T. (2007). *Truth and Ontology* (New York: Oxford University Press).

Miller, K. (2019). The Cresting Wave: A New Moving Spotlight Theory. *Canadian Journal of Philosophy*, 49(1): 94–122.

Miller, K., Holcombe, A., and Latham, A. J. (2018). Temporal Phenomenology: Phenomenological Illusion Versus Cognitive Error. *Synthese*, 197(2): 751–71.

Millikan, R. (1984). *Language, Thought and Other Biological Categories* (Cambridge, MA: MIT Press).

Millikan, R. (1989a). In Defense of Proper Functions. *Philosophy of Science*, 56(2): 288–302, and reprinted in Millikan, 1993(a) op. cit.

Millikan, R. (1989b). Biosemantics. *Journal of Philosophy*, 86: 281–97.

Monton, B. (2006). Presentism and Quantum Gravity. In *The Ontology of Spacetime*, D. Dieks (ed.) Elsevier.

Mumford, S. and Anjum, R. L. (2011). *Getting Causes from Powers* (OUP).

Nahmias, E., Morris, S., Nadelhoffer, T., and Turner, J. (2005). Surveying Freedom: Folk Intuitions About Free Will and Moral Responsibility. *Philosophical Psychology*, 18(5): 561–84.

Nahmias, E., Morris, S. G., Nadelhoffer, T., and Turner, J. (2006). Is Incompatibilism Intuitive? *Philosophy and Phenomenological Research*, 73(1): 28–53.

Nagel, Jennifer, Juan, V. S., and Mar, R. A. (2013). Lay Denial of Knowledge for Justified True Beliefs. *Cognition*, 129(3): 652–61. doi:10.1016/j.cognition.2013.02.008

Neander, K. (1991). Functions as Selected Effects. *Philosophy of Science*, 58: 168–84.

Neander, K. (1995). Malfunctioning and Misrepresenting. *Philosophical Studies*, 79: 109–41.

Neander, K. (1996). Swampman Meets Swampcow. *Mind and Language*, 11(1): 70–130.

Ney, A. (2012). The Status of Our Ordinary Three Dimensions in a Quantum Universe. *Nous*, 46(3): 525–60.

Ney, A. (2015). Fundamental Physical Ontologies and the Constraint of Empirical Coherence: A Defense of Wave Function Realism. *Synthese*, 192: 3105–24.

Nichols, S. and Knobe, J. (2007). Moral Responsibility and Determinism: The Cognitive Science of Folk Intuitions. *Noûs*, 41(4): 663–85.

Nishida, S. and Johnston, A. (1999). Influence of Motion Signals on the Perceived Position of Spatial Pattern. *Nature*, 18;297(6720): 610–2.

Nobre, A. C. and Coull, J. T. (2010). *Attention and Time* (Oxford: Oxford University Press).

Nolan, D., Restall, G., and West, C. (2005). Moral Fictionalism Versus the Rest. *Australasian Journal of Philosophy*, 83(3): 307–30.

Núñez, R. and Sweetser, E. (2006). With the Future Behind Them: Convergent Evidence from Aymara Language and Gesture in Crosslinguistic Comparison of Spatial Construals of Time. *Cognitive Science*, 30: 401–50.

Oaklander, L. N. (2012). A-, B-, and R-Theories of Time: A Debate. In B. Adrian (ed.), *The Future of the Philosophy of Time*, pp. 1–24 (NewYork: Routledge).

Olson, J. (2014). *Moral Error Theory: History, Critique, Defence* (Oxford: Oxford University Press).

Palmer, Stephen K. (1999). *Vision Science: Photons to Phenomenology* (Cambridge, MA: MIT Press).

Paoletti, P. (2016). A Sketch of (an Actually Serious) Meinongian Presentism. *Metaphysica*, 17(1): 1–18.

Paul, L. A. (2002). Logical Parts. *Noûs*, 36: 578–96.

Paul, L. A. (2010). Temporal Experience. *Journal of Philosophy*, 107(7): 333–59.

Pautz, Adam (2013). Does Phenomenology Ground Mental Content? In Uriah Kriegel (ed.), *Phenomenal Intentionality*, pp. 194–234 (Oxford: Oxford University Press).

Pezet, R. (2017). A Foundation for Presentism. *Synthese*, 194(5): 1809–37.

Phillips, I. (2010). Perceiving Temporal Properties. European *Journal of Philosophy*, 18(2): 176–202.

Phillips, I. (2011). Indiscriminability and Experience of Change. *Philosophical Quarterly*, 61 (245): 808–27.

Phillips, I. (2014). Breaking the Silence: Motion Silencing and Experience of Change. *Philosophical Studies*, 168(3): 693–707.

Pitt, David (2004). The Phenomenology of Cognition Or What Is It Like to Think That P? *Philosophy and Phenomenological Research*, 1: 1–36.

Polger, T. (2007). Realization and the Metaphysics of Mind. *Australasian Journal of Philosophy*, 85(2): 233–59.

Price, H. (1996). *Time's Arrow & Archimedes' Point: New Directions for the Physics of Time* (Oxford: Oxford University Press).

Price, H. (2007). Causal Perspectivalism. In Huw Price and Richard Corry (eds), *Causation, Physics and the Constitution of Reality: Russell's Republic Revisited* (Oxford: Oxford University Press).

Prior, A. (1967). *Past, Present and Future* (Oxford: Clarendon Press).

Prior, A. (1968). *Papers on Time and Tense* (Oxford: Oxford University Press).

Putnam, H. (1967). Time and Physical Geometry. *Journal of Philosophy*, 64: 240–47.

Quine, W. V. O. (1951). Two Dogmas of Empiricism. *Philosophical Review*, 60(1): 20–43.

Ramsey, Frank P. (1931). *The Foundations of Mathematics and Other Logical Essays* (Oxford: Routledge).

Rickles, D. (2011). A Philosopher Looks at String Dualities. *Studies in History and Philosophy of Modern Physics*, 42(1): 54–67.

Rosenkranz, S. and Correia, F. (2018). *Nothing to Come: A Defence of the Growing Block Theory of Time* (Springer Verlag).

Roskies, A. and Nichols, S. (2008). Bringing Moral Responsibility Down to Earth. *Journal of Philosophy*, 105(7): 371–88.

Rovelli, C. (1991). Time in Quantum Gravity: An Hypothesis. *Physical Review D*, 43(2): 451–6.

Rovelli, C. (2004). *Quantum Gravity* (Cambridge: Cambridge University Press).

Rovelli, C. (2007). The Disappearance of Space and Time. In Dennis Dieks (ed.), *The Ontology of Spacetime*, pp. 25–36 (Amsterdam: Elsevier).

Rovelli, Carlo and Vidotto, Francesca (2014). *Covariant Loop Quantum Gravity: An Elementary Introduction to Quantum Gravity and Spinfoam Theory* (Cambridge: Cambridge University Press).

Rudder-Baker, Lynn (2000). *Persons and Bodies: A Constitution View* (Cambridge: Cambridge University Press).

Saucedo, Raul (2011). Parthood and Location. In Karen Bennett and Dean Zimmerman (eds), *Oxford Studies in Metaphysics Vol. 6*, pp. 223–84 (Oxford: Oxford University Press).

Saunders, Simon (2013). Rethinking Newton's Principia. *Philosophy of Science*, 80(1): 22–48.

Schaffer, Jonathan (2009a). Spacetime the One Substance. *Philosophical Studies*, 145: 131–48.

Schaffer, Jonathan (2009b). On What Grounds What. In Manley, D., Chalmers, D., and Wasserman, R. (eds), *Metametaphysics*, pp. 347–83 (Oxford: Oxford University Press).

Schaffer, J. (2016). Grounding in the Image of Causation. *Philosophical Studies*, 173(1): 49–100.

Schlesinger, G. (1980). *Aspects of Time* (Indianapolis: Hackett).

Schlesinger, G. (1994). Temporal Becoming. In N Oakland and Q Smith (eds), *The New Theory of Time* (New Haven, CT: Yale University Press).

Schuster, M. M. (1986). Is the Flow of Time Subjective? *The Review of Metaphysics*, 39: 695–714.

Searle, John R. (1983). *Intentionality: An Essay in the Philosophy of Mind* (Cambridge: Cambridge University Press).

Searle, John R. (1990). *The Mystery of Consciousness* (Granta Books).

Searle, John R. (1992). *The Rediscovery of the Mind* (Cambridge, MA: MIT Press).

Shardlow, J., Lee, R., Hoerl, C., McCormack., T., Burns, P., and Fernandes, S. (2021). Exploring People's Beliefs About the Experience of Time. *Synthese*, 198: 10709–31.

Sider, Theodore (2007). Parthood. *Philosophical Review*, 116: 51–91.

Sider, T. (2013). Against Parthood. In K. Bennett and D. W. Zimmerman (eds), *Oxford Studies in Metaphysics*, Volume 8 (Oxford: Oxford University Press).

Simon, P. (1987). *Parts: A Study in Ontology* (Oxford: Oxford University Press).

Sinclair, N. (2012). Moral Realism, Face-Values and Presumptions. *Analytic Philosophy*, 53(2): 158–79.

Skow, B. (2011). Experience and the Passage of Time. *Philosophical Perspectives*, 25: 359–87.

Skow, B. (2015). *Objective Becoming* (Oxford: Oxford University Press).

Smith, Q. (1993). *Language and Time*. New York: (Oxford: Oxford University Press).

Smith, Q. (1994). The Phenomenology of a-Time. In L. Nathan Oaklander and Quentin Smith (eds), *The New Theory of Time*, pp. 351–9 (New Haven, London: Yale University Press).

Stampe, D. (1986). Verification and a Causal Account of Meaning. *Synthese*, 69: 107–37.

Steinman, R. M., Pizlo, Z., and Pizlo, F. J. (2000). Phi Is Not Beta, and Why Wertheimer's Discovery Launched the Gestalt Revolution. *Vision Research*, 40: 2257–64.

Stetson, C., Cui, X., Montague, P. R., and Eagleman, D. M. (2006). Motor-sensory Recalibration Leads To an Illusory Reversal of Action and Sensation. *Neuron*, 51.

Tallant, J. (2012). (Existence) Presentism and the A-theory. *Analysis*, 72(4): 673–81.

Tallant, J. (2018). An Error in Temporal Error Theory. *Journal of the American Philosophical Association*, 4(1): 14–32.

Tallant, J. and Ingram, D. (2020). A Defence of Lucretian Presentism. *Australasian Journal of Philosophy*, 98(4): 67–90.

Tegtmeier, E. (1996). The Direction of Time: A Problem of Ontology, not of Physics. In J. Faye (Hrsg.) *Perspectives on Time* (Dordrecht).

Tegtmeier, E. (2009). Ontology of Time and Hyperdynamism. *Metaphysica*, 10(2): 185–98.

Tegtmeier, E. (2014). Temporal Succession and Tense. In L. Nathan Oaklander (ed.), *Debates in the Metaphysics of Time*, pp. 73–86 (London: Bloomsbury).

Tegtmeier, E. (2016). Time and Order. *Manuscrito*, 39: 157–68.

Thunder, S. (forthcoming). Composite Objects Are Mere Manys. *Oxford Studies in Metaphysics*.

Tooley, M. (1997). *Time, Tense, and Causation* (Oxford University Press).

Torrengo, G. (2017a). The Myth of Presentism's Intuitive Appeal. 12: *New Trends in Philosophy*, https://doi.org/10.13128/Phe_Mi-21105

Torrengo, G. (2017b). Feeling the Passing of Time. *Journal of Philosophy*, cxiv(4): 165–88.

Tversky, B., Kugelmass, S., and Winter, A. (1991). Cross-Cultural and Developmental Trends In Graphic Productions. *Cognitive Psychology*, 23(4): 515–57. https://doi.org/10.1016/0010-0285(91)90005-9

Tye, Michael (2000). *Consciousness, Color and Content* (Cambridge, MA: MIT Press).

Tyler, C. W. (1973). Temporal Characteristics in Apparent Movement: Omega Movement vs. Phi Movement. *Quarterly Journal of Experimental Psychology*, 25: 182–92.

Walton, K. (1990). *Mimesis as Make-Believe: On the Foundations of the Representational Arts* (Harvard University Press).

Wasserman, Ryan (2004). The Constitution Question. *Noûs*, 38(4): 693–710.

Watzl, S. (2013). Silencing the Experience of Change. *Philosophical Studies*, 165(3): 1009–32.

Weatherall, James Owen (2016). Maxwell-Huygens, Newton-Cartan and Saunders-Knox Space-Times. *Philosophy of Science*, 83(1): 82–92.

Weinberg, Jonathan M., Nichols, Shaun, and Stich, Stephen (2001). Normativity and Epistemic Intuitions. *Philosophical Topics*, 29(1/2): 429–60. doi:10.5840/philtopics2001291/217

Williams, C. (1998). B-Time Transition. *Philosophical Inquiry*, 20(3/4): 59–63.

Williams, C. (2003). Beyond A- and B-Time. *Philosophia*, 31(1): 75–91.

Wilson, J. (2014). No Work for a Theory of Grounding. *Inquiry*, 57(5–6): 535–79.

Wilson, Jessica (2016). Metaphysical Emergence: Weak and Strong. *Metaphysics in Contemporary Physics*, Tomasz Bigaj and Christian Wüthrich (eds), pp. 345–402 (Leiden and Boston: Brill).

Wüthrich, Christian (2017). Raiders of the Lost Spacetime. In *Toward a Theory of Spacetime Theories*, D. Lehmkuhl, G. Schiemannn and E. Scholz (eds.) (Basal: Birkhäuser).

Wüthrich, Christian (2019). The Emergence of Space and Time. In *Routledge Handbook of Emergence*, Sophie Gibb, Robin Finlay Hendry and Tom Lancaster (eds.) (Oxford: Routledge).

Wüthrich, Christian and Callender, Craig (2017). What Becomes of a Causal Set? *British Journal for the Philosophy of Science*, 68(3): 907–25.

Yates, David (2021). Thinking About Spacetime. In Christian Wüthrich, Baptiste Le Bihan, and Nick Huggett (eds), *Philosophy Beyond Spacetime* (Oxford: Oxford University Press).

Zeki, S., (1983). Colour Coding in the Cerebral Cortex: The Reaction of Cells in Monkey Visual Cortex to Wavelengths and Colours. *Neuroscience*, 9(4): 741–81. doi:10.1016/0306-4522(83)90265-8

Zimmerman, D. W. (2008). The Privileged Present: Defending an 'A-Theory' of Time. In Sider, T., Hawthorne, J., and Zimmerman, D. W. (eds), *Contemporary Debates in Metaphysics*, 211–25 (Oxford: Blackwell).

Index